London Mathematical Society Lecture Note Series: 312

Foundations of Computational Mathematics: Minneapolis, 2002

Edited by

Felipe Cucker
The City University of Hong Kong

Ron DeVore
University of South Carolina

Peter Olver
University of Minnesota

Endre Süli
University of Oxford

PUBLISHED BY THE PRESS SYNDICATE OF THE UNIVERSITY OF CAMBRIDGE
The Pitt Building, Trumpington Street, Cambridge, United Kingdom

CAMBRIDGE UNIVERSITY PRESS
The Edinburgh Building, Cambridge CB2 2RU, UK
40 West 20th Street, New York, NY 10011–4211, USA
477 Williamstown Road, Port Melbourne, VIC 3207, Australia
Ruiz de Alarcón 13, 28014 Madrid, Spain
Dock House, The Waterfront, Cape Town 8001, South Africa

http://www.cambridge.org

© Cambridge University Press 2004

This book is in copyright. Subject to statutory exception
and to the provisions of relevant collective licensing agreements,
no reproduction of any part may take place without
the written permission of Cambridge University Press.

First published 2004

Printed in the United Kingdom at the University Press, Cambridge

Typeface Computer Modern 10/13pt   *System* LATEX 2_ε [AUTHOR]

A catalogue record for this book is available from the British Library

Library of Congress Cataloguing in Publication data available

ISBN 0 521 54253 7 paperback

LONDON MATHEMATICAL SOCIETY LECTURE NOTE SERIES

Managing Editor:
Professor N.J. Hitchin,
Mathematical Institute, 24–29 St. Giles, Oxford OX1 3DP, UK

All the titles listed below can be obtained from good booksellers or from Cambridge University Press. For a complete series listing visit http://publishing.cambridge.org/stm/mathematics/lmsn/

283. Nonlinear elasticity, R. OGDEN & Y. FU (eds)
284. Foundations of computational mathematics, R. DEVORE, A. ISERLES & E. SULI (eds)
285. Rational points on curves over finite fields, H. NIEDERREITER & C. XING
286. Clifford algebras and spinors, 2nd edn, P. LOUNESTO
287. Topics on Riemann surfaces and Fuchsian groups, E. BUJALANCE, A.F. COSTA & E. MARTINEZ (eds)
288. Surveys in combinatorics, 2001, J.W.P. HIRSCHFELD (ed)
289. Aspects of Sobolev-type inequalities, L. SALOFFE-COSTE
290. Quantum groups and Lie theory, A. PRESSLEY
291. Tits buildings and the model theory of groups, K. TENT
292. A quantum groups primer, S. MAJID
293. Second order partial differential equations in Hilbert spaces, G. DA PRATO & J. ZABCZYK
294. Introduction to operator space theory, G. PISIER
295. Geometry and integrability, L. MASON & Y. NUTKU (eds)
296. Lectures on invariant theory, I. DOLGACHEV
297. The homotopy theory of simply-connected 4-manifolds, H.J. BAUES
298. Higher operads, higher categories, T. LEINSTER
299. Kleinian groups and hyperbolic 3-manifolds, Y. KOMORI, V. MARKOVIC & C. SERIES (eds)
300. Introduction to Möbius differential geometry, U. HERTRICH-JEROMIN
301. Stable modules and the D(2)-problem, F.A.E. JOHNSON
302. Discrete and continuous nonlinear Schrödinger systems, M. ABLOWITZ, B. PRINARI & D. TRUBATCH
303. Number theory and algebraic geometry M. REID & A. SKOROBOGATOV
304. Groups St Andrews 2001 in Oxford vol. 1, C.M. CAMPBELL, E.F. ROBERTSON & G.C. SMITH (eds)
305. Groups St Andrews 2001 in Oxford vol. 2, C.M. CAMPBELL, E.F. ROBERTSON & G.C. SMITH (eds)
306. Peyresq lectures on geometric mechanics and symmetry, J. MONTALDI & T. RATIU (eds)
307. Surveys in combinatorics, 2003, C.D. WENSLEY (ed)
308. Topology, geometry and quantum field theory, U.L. TILLMANN (ed)
309. Corings and comodules, T. BRZEZINSKI & R. WISBAUER
310. Topics in dynamics and ergodic theory, S. BEZUGLYI & S. KOLYADA (eds)
311. Groups: topological, combinatorial and arithmetic aspects, T.W. MÜLLER (ed)

Contents

Preface		*page* vii
1	Some Fundamental Issues *R. DeVore*	1
2	Jacobi Sets *H. Edelsbrunner and J. Harer*	37
3	Approximation of boundary element operators *S. Börm and W. Hackbusch*	58
4	Quantum Complexity *S. Heinrich*	76
5	Straight-line Programs *T. Krick*	96
6	Numerical Solution of Structured Problems *T. Apel, V. Mehrmann, and D. Watkins*	137
7	Detecting Infeasibility *M.J. Todd*	157
8	Maple Packages and Java Applets *I.M. Anderson*	193

Preface

The Society for the Foundations of Computational Mathematics supports fundamental research in computational mathematics and its applications, interpreted in the broadest sense. As part of its endeavour to promote research across a wide spectrum of subjects concerned with computation, the Society regularly organises conferences and specialist workshops which bring together leading researchers working in diverse fields that impinge on various aspects of computation. Major conferences of the Society were held in Park City (1995), Rio de Janeiro (1997), Oxford (1999), and Minneapolis (2002).

The next FoCM conference will take place at the University of Santander in Spain in July 2005. More information about FoCM is available from the website `http://www.focm.net`.

The conference in Minneapolis on 5-14 August 2002 was attended by several hundred scientists. Workshops were held in eighteen fields which included: the foundations of the numerical solution of partial differential equations, geometric integration and computational mechanics, learning theory, optimization, special functions, approximation theory, computational algebraic geometry, computational number theory, multiresolution and adaptivity, numerical linear algebra, quantum computing, computational dynamics, geometrical modelling and animation, image and signal processing, stochastic computation, symbolic analysis, complexity and information-based complexity theory. In addition to the workshops, eighteen plenary lectures, concerned with a broad spectrum of topics connected to computational mathematics, were delivered by some of the world's foremost researchers. This volume is a collection of articles, based on the plenary talks presented at FoCM 2002. The topics covered in the lectures — ranging from the applications of computational mathematics in geometry and algebra to optimization theory, from quantum complexity to the numerical solution of partial differential equations, from numerical linear algebra to Morse theory — reflect the breadth of research within computational mathematics as well as the richness and fertility of interactions between seemingly unrelated branches of pure and applied mathematics.

We hope that the volume will be of interest to researchers in the field of computational mathematics but also to non-experts who wish to gain insight into the state of the art in this active and significant field.

Like previous FoCM conferences, the Minneapolis gathering proved itself as a unique meeting point of researchers in computational mathematics and of theoreticians in mathematics and in computer sciences. While presenting plenary talks by foremost world authorities and maintaining the highest technical level in the workshops, the emphasis, like in Park City, Rio de Janeiro and Oxford, was on multidisciplinary interaction across subjects and disciplines, in an informal and friendly atmosphere. It is only fair to say that for many of us the opportunity of meeting colleagues from different subject-areas and identifying the wide-ranging, and often surprising, common denominator to our research was a real journey of discovery.

We wish to express our gratitude to the local organisers and administrative staff of our hosts, the Institute of Mathematics and Its Applications and the Department of Mathematics at the University of Minnesota at Minneapolis, for making FoCM 2002 such a success. We also wish to thank the National Science Foundation, the Digital Technology Center in Minneapolis, IBM, the Office of Naval Research, the Number Theory Foundation and the American Institute of Mathematics for their generous sponsorship and support. Above all, however, we wish to express our gratitude to all participants of FoCM 2002 for attending the meeting and making it such an exciting, productive and scientifically stimulating event.

1

Some Fundamental Issues in Computational Mathematics

Ronald DeVore
Department of Mathematics
University of South Carolina
Columbia, SC 29208
Email: devore@math.sc.edu

Abstract

We enter a discussion as to what constitutes the 'foundations of computational mathematics'. While not giving a definition, we give examples from image/signal processing and numerical computation where foundational issues have helped to 'correctly' formulate problems and guide their solution.

1.1 The question

While past chair of the organization Foundations of Computational Mathematics (FOCM), I was frequently asked what is the meaning of 'foundations of computational mathematics'. Most people understand what computational mathematics is. So the question really centers around the meaning of 'foundations' in this context. Even though I have thought about this quite a while, I would not dare to try to give a precise definition of foundations – I am sure it would be picked apart. However, I would like in this presentation to give some examples where the adherence to fundamental questions has helped to shape the formulation of computational issues and more importantly contributed to their solution. The examples I choose in signal/image processing and numerical methods for PDEs are of course related to my own research. I am sure

---

[0] This work has been supported by the Office of Naval Research Contract Nr. N0014-91-J1343, the Army Research Office Contract Nr. DAAD 19-02-1-0028, and the National Science Foundation Grant DMS0221642.

there are many other stories of the type I put forward that are waiting to be told.

The first of the three examples that I will discuss is that of image compression. This subject has grown rapidly over the last decade with an important infusion of ideas from mathematics especially the theories of wavelets and nonlinear approximation. The main topic to be addressed here is how we can decide which algorithms for compression are optimal.

A somewhat related topic will concern Analog to Digital (A/D) conversion of signals. This is an area that is important in consumer electronics. The story here centers around trying to understand why engineers do A/D conversion in the way they do, which by the way is very counterintuitive to what a first mathematical analysis would suggest.

Finally, I discuss adaptive methods for solving PDEs. This is an extremely important area of numerical computation in our quest to solve large problems to higher and higher resolution. The question to be answered is how can we know when an adaptive method is optimal in its performance.

1.2 Image compression

Digital signal processing has revolutionized the storage and transmission of audio and video signals as well as still images, in consumer electronics and in more scientific settings (such as medical imaging). The main advantage of digital signal processing is its robustness: although all the operations have to be implemented with, of necessity, not quite ideal hardware, the a priori knowledge that all correct outcomes must lie in a very restricted set of well separated numbers makes it possible to recover them by rounding off appropriately.

Every day, millions of digitized images are created, stored, and transmitted over the Internet or using other mediums. A grey scale image is an array (matrix) of pixel values. It is already important to have a mathematical model for what these pixel values represent. We shall view the pixel array as arising in the following fashion. We have a light intensity function f defined on a continuum Ω. For simplicity we assume that $\Omega := [0,1]^2$ and that f takes values in $[0,1)$ (the latter can be achieved by simple renormalization). Digitization corresponds to two operations: averaging and quantization. We take a tiling of Ω into squares Q and associate to each square Q the average intensity

$$f_Q := \frac{1}{|Q|} \int_I f(x)\, dx,$$

where $|Q|$ denotes the Lebesgue measure of Q. The pixel values p_Q are derived from the numbers $f_Q \in [0,1)$ by quantization. We write f_Q in its binary expansion

$$f_Q = \sum_{j=1}^{\infty} b_j(f_Q) 2^{-j}$$

and define the pixel value $p_Q := \sum_{j=1}^{m} b_j(f_Q) 2^{-j}$. Typical choices of m are $m = 8$ (one byte per pixel) or $m = 16$. The array $I := I(f) := (p_Q)$ of pixel values is a digitization of f. The accuracy at which I resolves f depends on the fineness of the tiling and the accuracy of the quantization (i.e. size of m). We do not really know f. We only see it through the digitized image $I(f)$. In practice, the pixel values p_Q are corrupted by noise but we shall ignore this in our discussion since we are aiming in a different direction.

We see that a digitized image in its raw form is described by mN bits where N is the number of squares in the tiling. *Lossy compression* seeks to significantly reduce this number of bits used to represent f at the expense of some loss in the fidelity of resolution. Hopefully, this loss of fidelity is not perceptible. There are two parts to a lossy compression scheme. The *encoder* assigns to each pixel array I a bitstream $B(I)$. A *decoder* gives the recipe for changing any given bitstream B back into a pixel array. After encoding and then decoding the resulting pixel values $\bar{p}_Q$ will generally not be the same as the original p_Q. Some fidelity is lost.

One can imagine, given the practical importance of the compression problem, that there are a ton of encoding/decoding schemes. How can one decide from this myriad of choices which is the best? Engineers have introduced a number called the PSNR (Peak Signal to Noise Ratio) which measures the performance of a given encoding/decoding on a given digitized image I. It is not necessary to give its precise definition here but simply mention that it measures the least squares distortion $((\#I)^{-1} \sum_Q [p_Q - \bar{p}_Q]^2)^{1/2}$ as a function of the number of bits. Here $\#(I)$ is the number of pixels. A new encoding scheme is tested by its performance (PSNR) on a few test images – the Lena image being the most widely used.

Now, there is a fundamental question here. Should the quality of a compression algorithm be determined by its PSNR performance on a few test images? Given a collection of 2^k images, we can encode them all with k bits per image by simply enumerating them in binary. So on a mathematical level this test of performance is quite unsatisfactory.

What has any of this to do with "foundations of computational mathematics". Well, we cannot have a decidable competition among compression algorithms without a clear and precise formulation of the compression problem. This is a foundations question that rests on two issues that we must clarify. The first is the *metric* we are going to use to compare two images (for example, the original and the compressed image). The second is the class of images we wish to compress. We shall briefly discuss these issues.

1.2.1 The metric

We have already mentioned the PSNR which is based on the ℓ_2 metric. In our view of images as functions, this corresponds to the $L_2(\Omega)$ function metric. Is this the obvious choice? By no means. This choice seems to be more a matter of convenience and tradition. It is easy to solve optimization problems in the L_2 metric.

Certainly the choice of metric must depend on the intended application. In some targeted applications such as feature extraction and image registration, the least squares metric is clearly not appropriate and is better replaced by metrics such as L_∞ or maximum gradient.

Most compression is directed at producing visually pleasing images which cannot be distinguished from the original by the human eye. Thus, we can speak about the metric of the human visual system. The problem is that this vague notion is useless mathematically. Our goal would be to derive a mathematical metric which is a good model for the human visual system. There are some mathematical models for human vision which may be useful in directing our pursuit but little is agreed upon.

So, at this stage, we are left with using simple mathematical metrics such as the $L_p(\Omega)$ norms, $0 < p \leq \infty$, or certain smoothness norms. The point we wish to make here is not so much as to which metric is better but rather that any serious mathematical comparison of compression algorithms must at the outset agree on the metric used to measure distortion. Once this is decided we can go further.

1.2.2 Model classes of images

Once we have chosen a mathematical metric in which to measure distance between images, the question turns to describing the class of images that we wish to compress. This is a subject that spurs many interesting debates. We will touch on this only briefly and in a very prejudicial way.

There are two main models for images: the stochastic and the deterministic. Stochastic models are deeply embedded in the engineering and information theory communities influenced in a large part by Shannon's theory for optimal encoding. Deterministic models take the view we have presented of an image as a function defined on a continuum. We have begun by assuming only that an image is a bounded function. This is too broad of a class of functions to serve as the description of the images we wish to compress. Images have more structure.

One deterministic view of an image function f is that it is a sum of fundamental components corresponding to edges, texture, and noise. For example the famous model of Mumford and Shah [17] views the image as a sum $f = u + v$ of a component u of Bounded Variation (BV) and a component v in L_2. The component u is not an arbitrary BV function but rather has gradient given by a measure supported on a one dimensional set (corresponding to the edges in the image) and an L_1 part (corresponding to smooth regions in the image). The L_2 component v captures deviations from this model.

There are many variants of the Mumford–Shah model. These are beautifully described in the lecture notes of Meyer [14] – a must read. We wish to pick up on only one point of Meyer's exposition. Even when one settles on the functional nature of the two components u and v in the image, there are infinitely many ways to write $f = u + v$ depending on how much energy one wishes to put in each of these components. This is completely analogous to K-functional decompositions used in proving theorems on interpolation of operators. One needs to look at this totality of all such decompositions to truly understand f. For example, consider the case where we simply look for decompositions of $f = u + v$ where $u \in \mathrm{BV}$ and $v \in L_2$. We can give a quantitative description of these decompositions through the K-functional

$$K(f,t) := K(f,t; L_2, \mathrm{BV}) := \inf_{f = u+v} \|v\|_{L_2} + t|u|_{\mathrm{BV}}, \quad t > 0, \qquad (1.1)$$

where the $|\cdot|_{\mathrm{BV}}$ is the BV seminorm. For any fixed $t > 0$, the optimal decomposition in (1.1) tries to balance the two terms. Thus for t small it puts more energy into the BV component and less into the L_2 component. The rate of decrease in $K(f,t)$ as $t \to 0$ tells how nice f is with respect to this model.

The role of the K-functional is to distinguish between images. Certainly some images are more complex than others and more apt to be more difficult to compress. The rate at which a K-functional tends to

0 as $t \to 0$ measures this complexity. Thus, we can use the K-functional to separate images into classes K_α which are compact sets in our chosen metric. When classical metrics such as L_p norms are used, then these classes correspond to finite balls in smoothness spaces. In other words, using appropriate K-functionals, we can obtain a strata of image classes K_α reflecting the complexity of images.

1.2.3 Optimal encoding and Kolmogorov entropy

Suppose now that we have decided on the metric to be used to measure the distortion between two images and suppose we also have our model classes K_α for the classification of images. We shall assume that the metric is given by a quasi-norm $\|\cdot\| := \|\cdot\|_X$ on a topological linear space X. Each of the sets $K = K_\alpha$ is assumed to be a compact subset in the topology given by $\|\cdot\|$.

Recall that an encoder E for K is a mapping that sends each $f \in K$ into a bitstream $B(f) := B_E(f)$. Associated to E is a decoder D which takes any bitstream B and associates to it an element DB from X. Thus given $f \in K$, $\bar{f} := DEf = D(B_E(f))$ is the compressed image given by this encoding-decoding pair. This means that the distortion in the performance of this encoding on a given f is

$$d_E(f) := \|f - \bar{f}\| = \|f - DEf\|. \tag{1.2}$$

Of course, we are interested in the performance of this encoding not on just one element $f \in K$ but on the entire class. This leads us to define the *distortion for the class K* by

$$d_E(K) := \sup_{f \in K} d_E(f). \tag{1.3}$$

This distortion also depends on the decoder which we do not indicate. (One could become more specific here by always choosing for the given encoder E and set K the best decoder in the sense of minimizing the distortion (1.2).) To measure the complexity of the encoding we use

$$\#(E) := \#(E(K)) := \sup_{f \in K} \#(B(f)) \tag{1.4}$$

which is the maximum number of bits that E assigns to any of the elements of K.

We are interested in a competition among encoders/decoders to determine the optimal possible encoding of these classes. Suppose that we are given a bit budget n; this means we are willing to allocate a maximum of n bits in the encoding of any of the elements of K. Then,

$$d_n(K) := \inf_{\#(E) \leq n} d_E(K) \qquad (1.5)$$

is the minimal distortion that can be obtained for the class K with this bit budget n.

There is a mathematical description, called *Kolmogorov entropy*, that completely determines the optimal performance that is possible for an encoding of a given class K. Since K is compact in $\|\cdot\|$, for any given ϵ there is a collection of balls $\mathcal{B}(f_i, \epsilon)$, $i = 1, \ldots, N$, of radius ϵ centered at $f_i \in X$, such that

$$K \subset \bigcup_{i=1}^{N} \mathcal{B}(f_i, \epsilon). \qquad (1.6)$$

The smallest number $N_\epsilon(K)$ of balls that provide such a cover is called the *covering number* of K. The Kolmogorov entropy of K (in the topology of X) is then given by

$$H_\epsilon(K) := \log N_\epsilon(K) \qquad (1.7)$$

where here and later log always refers to the logarithm to the base 2. We fix K and think of $H_\epsilon(K)$ is a function of ϵ. It gives a measure of the massivity of K. The slower $H_\epsilon(K)$ tends to infinity as $\epsilon \to 0$ the more thin is the set K.

We can reverse the roles of ϵ and the entropy $H_\epsilon(K)$. Namely, given a positive integer n, let

$$\epsilon_n(K) := \inf\{\epsilon : H_\epsilon(K) \leq n\}. \qquad (1.8)$$

The $\epsilon_n(K)$ are called the entropy numbers of K; they tend to zero as $n \to \infty$. The faster they tend to zero, the smaller the set K. Notice that an asymptotic behavior $H_\epsilon(K) = \mathcal{O}(\epsilon^{-1/\alpha})$ is equivalent to $\epsilon_n(K) = \mathcal{O}(n^{-\alpha})$.

The two notions of optimal distortion and entropy numbers are identical:

$$\epsilon_n(K) = d_n(K). \qquad (1.9)$$

The proof is easy. Suppose E is an optimal encoder using n bits (if no such optimal encoder exists one modifies the following argument slightly). For each bitstream $B = B(f)$, $f \in K$, let $f_B := D(B)$ which is an element of X. Then, taking $\epsilon = d_n(K)$, we have $f \in \mathcal{B}(f_B, \epsilon)$. Since there are at most 2^n distinct bitstreams $B(f)$, $f \in K$, we obtain that $H_\epsilon \leq n$ and hence $\epsilon_n(K) \leq \epsilon = d_n(K)$. We can reverse this inequality as follows. Suppose that n is given and $\epsilon = \epsilon_n(K)$. We assume that $H_\epsilon(K) \leq n$. (We actually only know $H_\rho(K) \leq n$ for all $\rho > \epsilon$ so that our assumption is not necessarily valid but we can easily modify the argument when $\epsilon_n(K)$ is not attained.) Let $\mathcal{B}(f_i, \epsilon)$, $i = 1, \ldots, H_\epsilon(K)$, be a minimal covering for K with balls of radius ϵ. We associate to each i the binary bits in the binary representation of i. We define the encoder E as follows. If $f \in K$, we choose a ball $\mathcal{B}(f_i, \epsilon)$ that contains f (this is possible because these balls cover K) and we assign to f the bits in the binary representation of i. The decoder takes the bitstream, calculates the integer i which has these bits in its binary expansion, and assigns the decoded element to be the center of the ball $\mathcal{B}_i$. This encoding has distortion $\leq \epsilon = \epsilon_n(K)$ and so we have $d_n(K) \leq \epsilon_n(K)$.

The above discussion shows that the construction of an optimal encoder with distortion ϵ for the set K is the same as finding a minimal covering for K by balls of radius ϵ. Unfortunately, such coverings are usually impossible to find. For this reason, and others illuminated below, this approach is not very practical for encoding. On the other hand, it gives us a benchmark for the performance of encoders. If we could find an encoder which is nearly optimal for all the classes K of interest to us, then we could rest assured that we have done the job in the context in which we have framed the problem. We shall discuss in the next section how one could construct an encoder with these properties for a large collection of compact sets in standard metrics like L_p, $1 \leq p \leq \infty$.

1.2.4 Wavelet bases and compact subsets of L_p

A set K is compact in L_p provided that the modulus of smoothness

$$\omega(f, t)_p := \sup_{|h| \leq t} \|\Delta_h(f, \cdot)\|_{L_p(\Omega)}, \quad t > 0 \tag{1.10}$$

for all of the elements $f \in K$ have a continuous majorant ω_K

$$\sup_{f \in K} \omega(f, t)_p \leq \omega_K(t) \tag{1.11}$$

where $\omega_K(0) = 0$. The rate at which ω_K tends to zero at 0 measures the compactness of K. Thus the natural compact sets in L_p are described by common smoothness conditions. This leads to the Sobolev and Besov smoothness spaces. For example, the Besov spaces are defined by conditions on the higher order moduli of smoothness of $f \in L_p$. We denote these Besov spaces by $B_q^s(L_p(\Omega))$ where p is the L_p space in which we are measuring smoothness. The parameter $s > 0$ gives the order of smoothness much like the number of derivatives. The parameter $0 < q \leq \infty$ is a fine tuning parameter which makes subtle distinctions between these spaces. We do not make a precise description of these spaces at this juncture but we shall give a description of these spaces in a moment using wavelet bases.

The reader is probably by now quite familiar with wavelet bases. We shall limit ourselves to a few remarks which will serve to describe our notation. When working on the domain $\mathbb{R}$, a wavelet basis is given by the shifted dilates $\psi_\lambda := \psi(2^j \cdot -k)$, $\lambda = (j,k)$, of one fixed function ψ. When moving to $\mathbb{R}^d$, one needs the shifted dilates of a collection ψ^e of $2^d - 1$ functions; the parameter e is usually indexed on the set E of nonzero vertices of the unit cube $[0,1]^d$. Thus the wavelets are indexed by three parameters $\lambda = (j,k,e)$ indicated frequency (j), location (k) and type (e). When working on a finite domain, two adjustments need to be made. The first is that the range of j is from $j_0 \leq j < \infty$. The coarsest level $j = j_0$ corresponds to scaling functions; all other j correspond to the actual wavelets. For notational convenience, we shall take $j_0 = 0$ in what follows. The second adjustment is that near the boundary some massaging has to be made in defining ψ_λ.

Thus, a wavelet basis on a finite domain Ω in $\mathbb{R}^d$ is a collection $\Psi = \{\psi_\lambda : \lambda \in \mathcal{J}\}$ of functions ψ_λ. The indices λ encode scale, spatial location and the type of the wavelet ψ_λ. We will denote by $|\lambda|$ the *scale* associated with ψ_λ. We shall only consider compactly supported wavelets, i.e., the supports of the wavelets scale as follows

$$S_\lambda := \operatorname{supp} \psi_\lambda, \quad c_0 2^{-|\lambda|} \leq \operatorname{diam} S_\lambda \leq C_0 2^{-|\lambda|}, \qquad (1.12)$$

with c_0 and $C_0 > 0$ absolute constants. The index set $\mathcal{J}$ has the following structure $\mathcal{J} = \mathcal{J}_\phi \cup \mathcal{J}_\psi$ where $\mathcal{J}_\phi$ is finite and indexes the scaling functions on the fixed coarsest level 0. $\mathcal{J}_\psi$ indexes the "true wavelets" ψ_λ with $|\lambda| > 0$. From compactness of the supports we know that at each level, the set $\mathcal{J}_j := \{\lambda \in \mathcal{J} : |\lambda| = j\}$ is finite. In fact, one has $\#\mathcal{J}_j \sim 2^{jd}$ with constants depending on the underlying domain.

There is a natural tree structure associated to wavelet bases. A node in this tree corresponds to all $\lambda = (j,k,e)$, $e \in E$, with j,k fixed. In the case the domain is $\mathbb{R}^d$, each such node has 2^d children corresponding to the indices $(j+1, 2(k+e))$ where $e \in \{0,1\}^d$. In other words, the children all occur on the next dyadic level. In the case of Haar functions, the supports of the wavelets corresponding to children are contained in those corresponding to a given parent. This is modified on domains because only some of the indices are used on the domain.

Wavelet bases have many remarkable properties. The first that we want to pick up on is that Ψ is an *unconditional basis* for many function spaces X. Consider first the case that $X = L_2(\Omega)$. Then every $f \in L_2(\Omega)$ has a unique expansion $f = \sum f_\lambda \psi_\lambda$ and there exist some constants c and C independent of f such that

$$c\|(f_\lambda)_{\lambda \in \mathcal{J}}\|_{\ell_2} \leq \|\sum_{\lambda \in \mathcal{J}} f_\lambda \psi_\lambda\|_{L_2(\Omega)} \leq C\|(f_\lambda)_{\lambda \in \mathcal{J}}\|_{\ell_2}. \tag{1.13}$$

In the case of L_p spaces, $p \neq 2$, the norm $\|f\|_{L_p(\Omega)}$ is not so direct and must be made through the square function. However, if we normalize the basis in L_p, $\|\psi_\lambda\|_{L_p(\Omega)} = 1$, then the space B_p of functions $f = \sum_{\lambda \in \mathcal{J}} f_\lambda \psi_\lambda$ satisfying

$$\|f\|_{B_p} := \|(f_\lambda)\|_{\ell_p} \tag{1.14}$$

is very close to $L_p(\Omega)$ and can be used as a *poor man's* substitute in many instances. By the way, B_p is an example of a Besov space $B_p = B_p^0(L_p)$ where the smoothness order is zero.

Besov spaces in general have a simple description in terms of wavelet coefficients. If $f = \sum_{\lambda \in \mathcal{J}} f_\lambda \psi_\lambda$ with the ψ_λ normalized in L_p, $\|\psi_\lambda\|_{L_p(\Omega)} = 1$, then

$$\|h\|_{B_q^s(L_p(\Omega))} := \begin{cases} \left(\sum_{j=0}^\infty 2^{jsq} \left(\sum_{|\lambda|=j} |f_\lambda|^p\right)^{q/p}\right)^{1/q}, & 0 < q < \infty, \\ \sup_{j \geq 0} 2^{js} \left(\sum_{|\lambda|=j} |f_\lambda|^p\right)^{1/p}, & q = \infty. \end{cases} \tag{1.15}$$

Suppose that we fix $1 \leq p \leq \infty$ and agree to measure the distortion of images in the $L_p(\Omega)$ norm. Which of the Besov spaces are embedded in $L_p(\Omega)$ and which are compactly embedded? This is easily answered by the Sobolev embedding theorem. The unit ball of the Besov space $B_q^s(L_\tau(\Omega))$ is a compact subset if and only if $\frac{1}{\tau} < \frac{s}{d} - \frac{1}{p}$. Notice that this condition does not depend on q. When $\frac{1}{\tau} = \frac{s}{d} - \frac{1}{p}$ (the so-called *critical line* in the Sobolev embedding) then the Besov space $B_q^s(L_\tau(\Omega))$ is embedded in $L_p(\Omega)$ for small enough q but these embeddings are not compact.

1.2.5 Near optimal encoding in L_p, $1 \le p \le \infty$

Let us fix the metric of interest to us to be one of the L_p norms with $1 \le p \le \infty$. If we derive metrics that seem better measures of distortion for images then we would try to repeat the exercise of this section for these metrics.

We seek an encoder E with the following properties. First the encoder should be applicable to any function in L_p; such an encoder is said to be *universal*. We would like the encoder to give an infinite bitstream; giving more bits from the infinite bitsream would give a finer resolution of f. Such an encoder is called *progressive*. Finally, we would like the encoder to be near optimal for the compact sets that are unit balls of the Besov spaces in the following sense. If $E_n f$ denotes the first n bits of Ef, then we would like that for each such compact set K we have

$$\|D_n E_n f - f\|_{L_p(\Omega)} \le C_K \epsilon_n(K), \quad f \in K, \quad n = 1, 2, \ldots, \quad (1.16)$$

with C_K a constant depending only on K. This means that for these compact sets the encoder performs (save for the constant C_K) as well as any encoder.

It is quite amazing that there is a simple construction of encoders with these three desirable properties using wavelet decompositions. To describe these encoders, it is useful to keep in mind the case when the metric is L_2 and the wavelet basis is an orthonormal system. The general case follows the same principles but the proofs are not as transparent. Suppose then that $X = L_2(\Omega)$ and $f \in X$ with $\|f\|_X \le 1$ has an orthogonal wavelet expansion $f = \sum_{\lambda \in \mathcal{J}} f_\lambda \psi_\lambda$. To start the encoding, we would like to choose a few terms from this wavelet expansion which best represent f. The best choice we can make is to choose the largest terms since the remainder is always the sum of squares of the remaining coefficients.

Suppose $\eta > 0$ and $\Lambda_\eta := \Lambda_\eta(f) := \{\lambda : |f_\lambda| \ge \eta\}$ is the set of coefficients obtained by thresholding the wavelet coefficients at the threshold η. Note that $\Lambda_\eta = \emptyset$ when $\eta > 1$. We would like to encode the information about the set Λ_η and the coefficients f_λ, $\lambda \in \Lambda_\eta$. There are two issues to overcome. The first is that it is necessary to encode the positions of the indices in Λ_η. At first glance, these positions could occur anywhere which would cost possibly an arbitrarily large bit budget to encode them. But it turns out that for the sets K on which we want optimality of the encoding, namely K a unit ball of a Besov space, these positions align themselves at low frequencies. In fact, it can be proved [5] that whenever $f \in K$ one can find a tree T which contains Λ_η and is of comparable size to Λ_η. This motivates us to define for each $f \in L_2(\Omega)$, $T_\eta = T_\eta(f)$ as the smallest tree which contains $\Lambda_\eta(f)$. What we gain in going to a tree structure (which could be avoided at the expense of more complications and less elegance) is that it is easy to encode the positions of a tree using at most $2\#(T_\eta)$ bits. Indeed, one simply encodes the tree from its roots by assigning a bit 1 if the child of a current member λ of T_η is in T_η and zero otherwise; see [6, 5] for details.

The second issue to overcome is how to encode the coefficients f_λ for $\lambda \in T_\eta$. To encode just one real number we need an infinite number of bits. The way around this is the idea of quantization. In the present context, the (scalar) quantization is easy. Any real number y with $|y| \leq 1$ has a binary representation

$$y = (-1)^{s(y)} \sum_{i=0}^{\infty} b_i(y) 2^{-i} \tag{1.17}$$

with the $b_i(y) \in \{0, 1\}$ and the sign bit $s(y)$ defined as 0 if $y > 0$ and 1 otherwise. Receiving the bits $s(f), b_0(f), \ldots, b_m(f)$, we can approximate y by the partial sum $\bar{y} = (-1)^{s(y)} \sum_{i=0}^{m} b_i(y) 2^{-i}$ with accuracy

$$|y - \bar{y}| \leq 2^{-m}. \tag{1.18}$$

We apply this quantization to the coefficients f_λ, $\lambda \in T_\eta$. How should we choose m? Well in keeping with the strategy for thresholding, we would only want the residual $y - \bar{y}$ to be under the threshold η. Thus, if $\eta = 2^{-k}$, we would choose the quantization so that $m = k$.

There is only one other thing to note before we define our encoding. If η is a current threshold level and $\eta' < \eta$ is a new threshold then the tree $T_{\eta'}$ is a growing of the tree T_η. Keeping this in mind, let us take thresholds $\eta_k = 2^{-k}$, $k = 0, 1, \ldots$ and obtain the corresponding trees $T^k := T_{\eta_k}$, $k = 0, 1, \ldots$. To each $f \in L_2(\Omega)$, we assign the following bitstream:

$$P_0(f), S_0(f), B_0(f), P_1(f), S_1(f), B_1(f), B_{1,0}(f), \ldots \tag{1.19}$$

Here, $P_0(f)$ denotes the bits needed to encode the positions in the tree $T^0(f)$. The bits $S_0(f)$ are the sign bits of the coefficients corresponding to indices in $T^0(f)$. The set $B_0(f)$ gives the first bits $b_0(f_\lambda)$ of the coefficients f_λ corresponding to the $\lambda \in T^0$. When we advance k to the value 1, we assign in P_1 the bits needed to encode the new positions, i.e. the positions in $T^1 \setminus T^0$. Then $S_1(f)$ are the sign bits for the coefficients corresponding to these new positions. The bits $B_1(f)$ are the b_1 bits (which correspond to 2^{-1} in the binary expansion) of the coefficients corresponding to these new positions. Note that each new coefficient has absolute value $\leq 1/2$ so the bit $b_0 = 0$ for these coefficients. The set $B_{1,0}(f)$ gives the second bit (i.e. the b_1 bit) in the binary expansion of the f_λ, $\lambda \in T^0$. The reason we add these bits is so that each coefficient, whether it is from T^0 or T^1 is resolved to the same accuracy (i.e. accuracy $1/2$ at this stage). The process continues as we increase k. We send position bits to identify the new positions, a sign bit and a lead bit for each coefficient corresponding to a new position, and then one bit from the binary expansion of all the old positions.

We denote by E the mapping which takes a given $f \in L_2(\Omega)$ into the bitstream (1.19). For each $k \geq 0$, we let E_k be the encoder obtained from E by truncating at stage k. Thus $E_k(f)$ is the finite bitstream

$$P_0(f), S_0(f), B_0(f), \ldots, P_k(f), S_k(f), B_k(f), B_{k,0}(f), \ldots, B_{k,k-1}(f)$$
(1.20)

Let us say a few words about the decoding of such a bitstream. When a receiver obtains a bitstream of the form (1.20), he knows the first bits will give the positions of a tree of wavelet indices. From the form of the tree encoding, he will know when the bits of P_0 have ended; see [5]. At this stage he knows the number of elements in the tree T^0. He therefore knows the next $\#(T^0)$ bits will give the signs of the corresponding coefficients and the following $\#(T^0)$ bits will be the binary bits for position 0 in the binary expansion of each of these coefficients. Then, the process repeats itself at level 1.

The encoder E has all of the properties we want. It is universal, i.e. defined for each $f \in L_2(\Omega)$. It is progressive: as we receive more bits we get a finer resolution of f. Finally it is near optimal in the following sense. If K is the unit ball of any of the Besov spaces which are compactly embedded into $L_2(\Omega)$, then for any k the encoder E_k is near optimal in the sense of (1.16).

While we have discussed the encoder E in the context of measuring distortion in L_2, the same ideas apply when distortion is measured in L_p for any $1 \leq p \leq \infty$. The only alteration necessary is to work with the wavelet decomposition with wavelets normalized in $L_p(\Omega)$ which of course alters the coefficients as well.

1.3 Analog to Digital (A/D) conversion

As we have previously observed, the digital format is preferred for representing signals because of its robustness. However, many signals are not digital but rather analog in nature; audio signals, for instance, correspond to functions $f(t)$, modeling rapid pressure oscillations, which depend on the "continuous" time t (i.e. t ranges over $\mathbb{R}$ or an interval in $\mathbb{R}$, and not over a discrete set), and the range of f typically also fills an interval in $\mathbb{R}$. For this reason, the first step in any digital processing of such signals must consist in a conversion of the analog signal to the digital world, usually abbreviated as A/D conversion. Note that at the end of the chain, after the signal has been processed, stored, retrieved, transmitted,..., all in digital form, it needs to be reconverted to an analog signal that can be understood by a human hearing system; we thus need a D/A conversion there.

There are many proposed algorithms for A/D conversion. As in our discussion the last section, we would like to understand how we could decide which of these algorithms is optimal for encoding/decoding. As in that case, we have two initial issues: determine the metric to be used to measure distortion and the class of signals that are to be encoded. The metric issue is quite similar to that for images except now the human visual system is replaced by the human auditory system. In fact such considerations definitely play a role in the design of good algorithms but as of yet we are aware of no mathematical metric which is used to model the auditory system in the mathematical analysis of the encoding problem. The two metrics usually utilized in distortion analysis are the $L_2(\mathbb{R})$ and $L_\infty(\mathbb{R})$ norms.

Concerning model classes for auditory signals, it is customary to model audio signals by *bandlimited* functions, i.e. functions $f \in L^2(\mathbb{R})$ for which the Fourier transform

$$\hat{f}(\xi) = \frac{1}{\sqrt{2\pi}} \int_{-\infty}^{\infty} f(t) e^{-i\xi t} dt$$

vanishes outside an interval $|\xi| \leq \Omega$. The bandlimited model is justified by the observation that for the audio signals of interest to us, observed over realistic intervals time intervals $[-T, T]$, $\|\chi_{|\xi|>\Omega}(\chi_{|t|\leq T} f)^\wedge\|_2$ is negligible compared with $\|\chi_{|\xi|\leq\Omega}(\chi_{|t|\leq T} f)^\wedge\|_2$ for $\Omega \simeq 2\pi \cdot 20,000$ Hz.

So in proceeding further, let us agree that we shall use either the L_2 or L_∞ metric and the class of functions we shall consider are those that are bounded and bandlimited. We define $\mathcal{S}$ to be the collection of all functions in $L_2 \cap L_\infty$ whose L_2 norm is ≤ 1 and L_∞ norm is $\leq a < 1$ with $a > 0$ fixed and whose Fourier transform vanishes outside of $[-\pi, \pi]$. The choice of a and π are arbitrary; indeed given any $f \in L_2 \cap L_\infty$ which is bandlimited, we can dilate f and then multiply by a constant to arrive at an element of $\mathcal{S}$. Thus any encoders derived for $\mathcal{S}$ can easily be applied to general f.

The class $\mathcal{S}$ has a lot of structure. There is a well-known sampling theorem that says that any function $f \in \mathcal{S}$ is completely determined by its values on $\mathbb{Z}$. Indeed, we can recover f from the formula

$$f(t) = \sum_{n \in \mathbb{Z}} f(n) \frac{\sin(t-n)}{(t-n)} = \sum_{n \in \mathbb{Z}} f(n) \mathrm{sinc}(t-n) . \tag{1.21}$$

This formula is usually referred to as the Shannon-Whitaker formula. The sampling rate of 1 (called the Nyquist rate) arises because $\hat{f}$ vanishes outside of $[-\pi, \pi]$. The sinc functions appearing on the right side of (1.21) form an orthonormal system in L_2. Changing the support interval from $[-\pi, \pi]$ to $[-A\pi, A\pi]$ would correspond to Nyquist sampling rate of $1/A$.

The proof of (1.21) is simple and instructive. We can write

$$\hat{f} = F \cdot \chi \tag{1.22}$$

where $F := \sum_{k \in \mathbb{Z}} \hat{f}(\cdot + 2k\pi)$ is the periodization of $\hat{f}$ and χ is the characteristic function of $[-\pi, \pi]$. The Fourier coefficients of F are $\hat{F}(n) = f(n)$ and so $F = \sum_{n \in \mathbb{Z}} f(n) e^{in\omega}$. Substituting that into (1.22) and inverting the Fourier transform we obtain (1.21) because the inverse Fourier transform of χ is the sinc function.

With the formula (1.21) in hand, it seems that our quest for an optimal encoder is a no brainer. We should simply quantize the Nyquist samples $f(n)$. Given a real number $y \in (-1, 1)$, we can write as before

$$y = (-1)^{s(y)} \sum_{i=1}^{\infty} b_i(y) 2^{-i} \tag{1.23}$$

where $(-1)^{s(y)}$, $s(y) \in \{0, 1\}$, is the sign of y and the b_i are the binary bits of $|y|$. If we pick a number $m > 0$, the quantized values

$$\bar{y} = (-1)^{s(y)} \sum_{i=1}^{m} b_i(y) 2^{-i} \tag{1.24}$$

can be described by m bits and $|y - \bar{y}| \leq 2^{-m}$. If we apply this to the samples $f(n)$, $n \in \mathbb{Z}$, we have an encoding of f that uses m bits per Nyquist sample. Encoders built on this simple idea are called Pulse Code Modulation (PCM). However, they are not the encoders of choice in A/D conversion. Our excursion into this topic in this section is to understand why this is the case. Can we explain mathematically why engineers do not prefer PCM and better yet to explain the advantages of what they do prefer.

To begin the story, we have to dig a bit deeper into what we really mean by an encoding of a signal. The formula (1.21) requires an infinite number of samples to recover f and therefore apparently an infinite number of bits. Of course, we cannot compute, store, or transmit an infinite bitstream. But fortunately, we only want to recover f on a finite time interval which we shall take to be $[0, T]$. Even then, the contribution of samples far away from $[0, T]$ is large. Indeed, if we incur a fixed error δ in representing each sample, then the total error on $[0, T]$ is possibly infinite since the sinc functions decay so slowly: $\sum_{n \in \mathbb{Z}} |\text{sinc}(t-n)| = \infty$.

There is a way around this by sampling the function f at a slightly higher rate. Let $\lambda > 1$ and let g_λ be a C^∞ function such that $\hat{g}_\lambda = 1$ on $[-\pi, \pi]$ and $\hat{g}_\lambda$ vanishes outside of $[-\lambda\pi, \lambda\pi]$. Returning to our derivation of (1.21), using $\hat{g}_\lambda$ in place of χ, we obtain the representation

$$f(t) = \frac{1}{\lambda} \sum_{n \in \mathbb{Z}} f\left(\frac{n}{\lambda}\right) g_\lambda\left(t - \frac{n}{\lambda}\right). \tag{1.25}$$

Because g is smooth with fast decay, this series now converges absolutely and uniformly; moreover if the $f\left(\frac{n}{\lambda}\right)$ is replaced by $\widetilde{f}_n = f\left(\frac{n}{\lambda}\right) + \varepsilon_n$ in (1.25), with $|\varepsilon_n| < \varepsilon$, then the difference between the approximation $\widetilde{f}(x)$ and $f(x)$ can be bounded uniformly:

$$|f(t) - \widetilde{f}(t)| \leq \varepsilon \frac{1}{\lambda} \sum_{n \in \mathbb{Z}} \left|g\left(t - \frac{n}{\lambda}\right)\right| \leq \varepsilon C_g \tag{1.26}$$

where $C_g = \lambda^{-1} \|g'\|_{L^1} + \|g\|_{L^1}$ does not depend on T. Oversampling thus buys the freedom of using reconstruction formulas, like (1.25), that weigh the different samples in a much more localized way than (1.21) (only the $f\left(\frac{n}{\lambda}\right)$ with $\left|t - \frac{n}{\lambda}\right|$ "small" contribute significantly). In practice, it is customary to sample audio signals at a rate that is about 10 or 20% higher than the Nyquist rate; for high quality audio, a traditional sampling rate is 44,000 Hz.

One can show that the above idea of sampling slightly higher than the Nyquist rate and then quantizing the samples using binary expansion peforms at the Kolmogorov entropy rate for the class $\mathcal{S}$ (at least when T is large). This was first proved in [13] and later repeated in [10]. We spare the details and refer the reader to either of these two papers. So it seems now that the matter of encoding is closed by using such a modified PCM encoding. But I have surprising news: engineers do not prefer this method in practice. In fact they prefer another class of encoders known as Sigma Delta Modulation which we shall describe below. The mystery we still want to uncover is why they prefer these methods.

To explain the Sigma-Delta story, we return to the idea of oversampling which is at the heart of these encoders. We have seen the benefits of slight oversampling: sampling slightly higher than the Nyquist rate and giving several bits for each sample performs at Kolmogorov entropy rates. Sigma-Delta encoders go to the other extreme. They sample at a rate $\lambda > 1$ which is very large but then allot only one bit to each sample. Thus, to each sample $f(\frac{n}{\lambda})$ they assign a single bit $q_n^\lambda \in \{-1, 1\}$. On the surface this is very counter intuitive. If we think of that one bit as giving an approximation to the sample $f(\frac{n}{\lambda})$ then we cannot do very well. In fact the best we can do is give the sign of the sample. But the Sigma-Delta encoders do not do this. Rather, they make their bit assignment to $f(\frac{n}{\lambda})$ based on the past samples $f(\frac{m}{\lambda})$, $m < n$, and the bits that have already been assigned to them.

Let us describe this in more detail by considering the simplest of these encoders. We introduce an auxiliary sequence $(u_n)_{n \in \mathbb{Z}}$ (sometimes described as giving the "internal state" of the Sigma-Delta encoder) iteratively defined by

$$\begin{cases} u_n = u_{n-1} + f\left(\frac{n}{\lambda}\right) - q_n^\lambda \\ q_n^\lambda = \text{sign}\left(u_{n-1} + f\left(\frac{n}{\lambda}\right)\right), \end{cases} \quad (1.27)$$

and with an "initial condition" $u_0 = 0$. The q_n^λ are the single bit we assign to each sample. In circuit implementation, the range of n in (1.27) is $n \geq 1$. However, for theoretical reasons, we view (1.27) as defining the u_n and q_n for all n. At first glance, this means the u_n are defined implicitly for $n < 0$. However, it is possible to write u_n and q_n directly in terms of u_{n+1} and f_{n+1} when $n < 0$; see [9].

The role of the auxiliary sequence (u_n) is to track the difference between the running sums of the $f(\frac{n}{\lambda})$ and those of the q_n^λ. It is easy to see that the choice for q_n^λ used in (1.27) keeps the difference of these running sums to be ≤ 1. For this, one proves simply by induction that the $|u_n| \leq 1$, for all $n \in \mathbb{Z}$.

Of course, we need to describe how we decode the bit stream q_n^λ. For this we use (1.25) with $f(\frac{n}{\lambda})$ replaced by q_n^λ:

$$\bar{f} := \frac{1}{\lambda} \sum_{n \in \mathbb{Z}} q_n^\lambda g_\lambda(n - \frac{n}{\lambda}). \quad (1.28)$$

At this point, we have no information about the accuracy at which $\bar{f}$ represents f. However, simple estimates are available using summation by parts. For any $t \in \mathbb{R}$, we have

$$\left| f(t) - \frac{1}{\lambda} \sum_n q_n^\lambda g_\lambda \left(t - \frac{n}{\lambda} \right) \right|$$

$$= \frac{1}{\lambda} \left| \sum_n \left(f\left(\frac{n}{\lambda}\right) - q_n^\lambda \right) g_\lambda \left(t - \frac{n}{\lambda} \right) \right|$$

$$= \frac{1}{\lambda} \left| \sum_n u_n \left(g_\lambda \left(t - \frac{n}{\lambda} \right) - g_\lambda \left(t - \frac{n+1}{\lambda} \right) \right) \right|$$

$$\leq \frac{1}{\lambda} \sum_n \left| g_\lambda \left(t - \frac{n}{\lambda} \right) - g_\lambda \left(t - \frac{n+1}{\lambda} \right) \right|$$

$$\leq \frac{1}{\lambda} \sum_n \int_{t-\frac{n+1}{\lambda}}^{t-\frac{n}{\lambda}} |g_\lambda'(y)| dy = \frac{1}{\lambda} \|g_\lambda'\|_{L^1} \leq \frac{C}{\lambda}.$$

where we have used the fact that $|u_n| \leq 1$, for all $n \in \mathbb{Z}$.

There is good news and bad news in the last estimate. The good news is that we see that $\bar{f}$ approximates f better and better as the sampling rate λ increases. The bad news is that this decay $\mathcal{O}(\frac{1}{\lambda})$ is far inferior to the exponential rate provided by PCM. Indeed, for an investment of m bits per Nyquist sample PCM provides distortion $\mathcal{O}(2^{-m})$ whereas for an investment of λ bits per Nyquist sample, Sigma-Delta provides distortion $\mathcal{O}(1/\lambda)$. So at this stage, we still have no clue why Sigma-Delta Modulators are preferred in practice.

It is possible to improve the rate distortion for Sigma-Delta encoding by using higher order methods. In [9], a family of such encoders were constructed. The encoder of order k is proven to give rate distortion $\mathcal{O}(\lambda^{-k})$. If one allows k to depend on λ, one can derive error bounds of the order $\mathcal{O}(2^{-(\log \lambda)^2})$ – still far short of the exponential decay of PCM. The pursuit of higher performing encoders has led to a series of interesting questions concerning the optimal distortion possible using single bit encoders. The best known bound $(e^{-.07\lambda})$ for such encoders was given by Güntürk [12], Calderbank and Daubechies [4] show that it is not possible to obtain the rate $(2^{-\lambda})$ of PCM. In any case, none of these new methods are used in practical encoding and they do not explain the penchant for the classical Sigma-Delta methods.

A couple of years ago, Ingrid Daubechies, Sinan Güntürk, and I were guests of Vinay Vaishampayan at Shannon Labs of AT&T for a one

1. Some Fundamental Issues

month think tank directed at understanding the preferences for Sigma-Delta Modulation. Shannon Labs is an oasis for Digital Signal Processing and its circuit implementation. We were fortunate to have many lunch with experts in Sigma-Delta methods asking them for their intuition why these methods are preferred in practice. This would be followed by an afternoon to put a mathematical justification behind their ideas. Usually, these exercises ended in futility but what became eventually clear is that the circuit implementation of Sigma-Delta Modulation is at the heart of the matter.

The hardware implementation of encoders such as PCM or Sigma-Delta Modulation requires building circuits for the various mathematical operations and the application of quantizers Q such as the sign function used in (1.27). Our mathematical analysis thus far has assumed that these operations are made exactly. Of course, this is far from the case in circuit implementation.

Let us suppose for example that the *sign* function used in (1.27) is replaced at iteration n by a non-ideal quantizer

$$\begin{aligned} Q_n(x) &= \text{sign}(x) \text{ for } |x| \geq \tau \\ |Q_n(x)| &\leq 1 \qquad \text{for } |x| < \tau. \end{aligned} \qquad (1.29)$$

Thus, the quantizer Q_n gives the right value of sign when $|x| \geq \tau$ but may not when $|x| < \tau$. Here τ would depend on the precision of the circuit which of course is related to the dollar investment we want to make in manufacturing such circuits. Note that we allow the quantization to vary with each application but require the overall precision τ. When we implement the Sigma-Delta recursion with this circuitry, we would not compute the q_n^λ of (1.27) but rather a new sequence $\bar{q}_n$ given by

$$\begin{cases} \bar{u}_n = \bar{u}_{n-1} + f\left(\frac{n}{\lambda}\right) - \bar{q}_n \\ \bar{q}_n = Q_n\left(\bar{u}_{n-1} + f\left(\frac{n}{\lambda}\right)\right), \end{cases} \qquad (1.30)$$

Thus, the error analysis given above is sort of irrelevant and needs to be replaced by one involving the $\bar{q}_n$. Recall that our analysis above rested on showing the state variables u_n are uniformly bounded. It turns out that in the scenario the new state variables $\bar{u}_n$ are then still bounded, uniformly, independently of the detailed behavior of Q_n, as long as (1.29) is satisfied. Namely, we have

Remark 1.1 *Let $f \in \mathcal{S}$, let $\bar{u}_n, \bar{q}_n$ be as defined in (1.30), and let Q_n satisfy (1.29) for all n. Then $|\bar{u}_n| \leq 1 + \tau$ for all $n \geq 0$.*

We refer the reader to [9] for the simple proof. Note that the remark holds regardless of how large τ is; even $\tau \gg 1$ is allowed.

We now use the inaccurate bits $\bar{q}_n$ to calculate $\bar{f}$: The same summation by parts argument that derived (1.29) can be applied to derive the new error estimate:

$$|f(t) - \bar{f}(t)| \leq \frac{(1+\tau)\|g'_\lambda\|_{L_1}}{\lambda}. \tag{1.31}$$

Thus, except for the fact that the constants increase slightly, the bounds on the accuracy of the encoder does not change. The precision that can be attained is not limited by the circuit imperfection: by choosing λ sufficiently large, the approximation error can be made arbitrarily small.

The same is definitely not true for the binary expansion-type schemes such as PCM. To see this, let us see how we quantize to obtain the binary bits of a number $y \in (-1, 1)$. To find the sign bit of y, we use the quantizer Q as before. But to find the remaining bits, we would use

$$Q_1(z) := \begin{cases} 0, & z \leq 1 \\ 1, & z > 1. \end{cases} \tag{1.32}$$

Once the sign bit b_0 is found then, we define $u_1 := 2b_0 y = 2|y|$. The bit b_1 is given by $b_1 := Q_1(u_1)$. Then the remaining bits are computed recursively as follows: if u_i and b_i have been defined, we let

$$u_{i+1} := 2(u_i - b_i) \tag{1.33}$$

and

$$b_{i+1} := Q_1(u_{i+1}). \tag{1.34}$$

In circuit implementations, the quantization would not be exact. Suppose for example, we use an imprecise quantizer (1.29) to find the sign bit of y. Then, taking a $y \in (-\tau, \tau)$, we may have the sign bit of y incorrect. Therefore, $\bar{y}$ and y do not even agree in sign so $|\bar{y} - y|$ could be as large as τ no matter how the remaining bits are computed. The mistake made by the imperfect quantizer cannot be recovered by computing more bits, in contrast to the self-correcting property of the Sigma-Delta scheme.

So there it is! This is a definite advantage in Sigma-Delta Modulation when compared with PCM. In order to obtain good precision overall with the binary quantizer, one must therefore impose very strict requirements on τ, which would make such quantizers very expensive in practice (or even impossible if τ is too small). On the other hand [10], Sigma-Delta encoders are robust under such imperfections of the quantizer, allowing for good precision even if cheap quantizers are used (corresponding to less stringent restrictions on τ). It is our understanding that it is this feature that makes Sigma-Delta schemes so successful in practice.

We have shown again where the understanding and formulation of fundamental questions in computation is vital to understanding numerical methods. In this case of A/D conversion, it not only gives an understanding of the advantages of the current state of the art encoders, it also leads us to a myriad of questions at the heart of the matter [9, 10, 12]. We shall pick up on just one of these.

We have seen that oversampling and one bit quantization allow error correction in the encoding but the bit distortion rate in the encoding is not very good. On the other hand, PCM has excellent distortion rate (exponential) but no error correction. It is natural to ask whether we can have the best of both worlds: exponential decay in distortion and error correction. The key to answering this question lies in the world of redundancy. We have seen the effect of the redundancy in the Sigma-Delta Modulation which allowed for error correction. It turns out that other types of redundancy can be utilized in PCM encoders. The essential idea is to replace the binary representation of a real number y by a redundant representation.

Let $1 < \beta < 2$ and $\gamma := 1/\beta$. Then each $y \in [0,1]$ has a representation

$$y = \sum_{i=1}^{\infty} b_i \gamma^i \tag{1.35}$$

with

$$b_i \in \{0, 1\}. \tag{1.36}$$

In fact there are many such representations. The main observation that we shall utilize below is that no matter what bits b_i, $i = 1, \ldots, m$, have been assigned, then, as long as

$$y - \frac{\gamma^{m+1}}{1-\gamma} \leq \sum_{i=1}^{m} b_i \gamma^i \leq y, \tag{1.37}$$

there is a bit assignment $(b_k)_{k>m}$, which, when used with the previously assigned bits, will exactly recover y.

We shall use this observation in an analogous fashion to the algorithm for finding the binary bits of real numbers, with the added feature of quantization error correction.

These encoders have a certain offset parameter μ whose purpose is to make sure that even when there is an imprecise implementation of the encoder, the bits assigned will satisfy (1.37); as shown below, introducing μ corresponds to carrying out the decision to set a bit to 1 only when the input is well past its minimum threshold. We let Q_1 be the quantizer of (1.32).

The beta-encoder with offset μ. Let $\mu > 0$ and $1 < \beta < 2$. For $y \in [0,1]$, we define $u_1 := \beta y$ and $b_1 := Q_1(u_1 - \mu)$. In general, if u_i and b_i have been defined, we let

$$u_{i+1} := \beta(u_i - b_i), \quad b_{i+1} := Q_1(u_{i+1} - \mu). \tag{1.38}$$

It then follows that

$$y - \sum_{i=1}^{m} b_i \gamma^i = y - \sum_{i=1}^{m} \gamma^i (u_i - \gamma u_{i+1})$$
$$= y - \gamma u_1 + \gamma^{m+1} u_{m+1} \leq \gamma^{m+1} \|u\|_{l^\infty}, \tag{1.39}$$

showing that we have exponential precision in our reconstruction, provided the $|u_i|$ are uniformly bounded. It is easy to prove [10] that we do indeed have such a uniform bound. Let's analyze the error correcting abilities of these encoders when the quantization is imprecise.

Suppose that in place of the quantizer Q_1, we use at each iteration in the beta-encoder the imprecise quantizer

$$\tilde{Q}_1(z) := \begin{cases} 0, & z \leq 1 - \tau \\ 1, & z > 1 + \tau \\ \in \{0, 1\}, & z \in (-\tau, \tau). \end{cases} \tag{1.40}$$

In place of the bits $b_i(y)$, we shall obtain inaccurate bits $\tilde{b}_i(y)$ which are defined recursively by $\tilde{u}_1 := \beta y$, $\tilde{b}_1 := \tilde{Q}_1(\tilde{u}_1 - \mu)$ and more generally,

$$\tilde{u}_{i+1} := \beta(\tilde{u}_i - \tilde{b}_i), \quad \tilde{b}_{i+1} := \tilde{Q}_1(\tilde{u}_{i+1} - \mu). \tag{1.41}$$

Theorem 1.1 *Let $\delta > 0$ and $y \in [0,1)$. Suppose that in the beta-encoding of y, the quantizer $\tilde{Q}_1$ is used in place of Q_1 at each occurrence, with the values of τ possibly varying but always satisfying $|\tau| \leq \delta$. If $\mu \geq \delta$ and β satisfies*

$$1 < \beta \leq \frac{2 + \mu + \delta}{1 + \mu + \delta}, \tag{1.42}$$

then for each $m \geq 1$, $\tilde{y}_m := \sum_{k=1}^{m} \tilde{b}_k \gamma^k$ satisfies

$$|y - \tilde{y}_m| \leq C\gamma^m, \quad m = 1, 2, \ldots, \tag{1.43}$$

with $C = 1 + \mu + \delta$.

The simple proof of this theorem is given in [10].

The beta encoder can be used in place of binary encoder in a PCM type encoding. We sample at slightly higher than Nyquist rate, i.e., λ is slightly larger than one. For each sample $f(n/\lambda)$, we use the beta encoder to determine m bits in the beta expansion of this sample. The corresponding bitstream will therefore assign slightly more than m bits per Nyquist sample. Decoding these bits gives an approximation $\bar{f}_n$ to $f(n/\lambda)$. Even if the quantization is not exact, we will have the accuracy

$$|f(n/\lambda) - \bar{f}_n| \leq C\beta^m.$$

Therefore, the signal $\bar{f}$ reconstructed using these bits will have exponential accuracy as well:

$$|f(t) - \bar{f}(t)| \leq C\beta^m, \quad t \in \mathbb{R}. \tag{1.44}$$

1.4 Adaptive methods for PDEs

The third and last topic we wish to engage concerns the numerical computation of solutions to PDEs. Given such an equation, how can we decide if a given numerical method is best possible? We shall see that there are three intertwining ingredients here: approximation theory, regularity theorems for PDEs, and analysis of the given numerical method.

Any numerical method can be viewed as a form of approximation. In classical Finite Element Methods (FEM), the approximation tool is piecewise polynomials subordinate to partitions of the domain for the PDE. The role of approximation theory is to tell us what we can expect as a best performance using such an approximation tool. It assumes that the solution is known to us in its analysis of performance and therefore does not apply directly to our unknown solution. The usual form of an approximation theorem is to characterize precisely which functions are approximated with a specified order by the approximation tool. For example, if n is the number of parameters used in the approximation, then the typical theorem says that a function can be (best) approximated with an error $\mathcal{O}(n^{-s})$ if and only if f is in a certain smoothness space $\mathcal{A}^s$.

To use such approximation theorems as a gauge of the performance of the numerical method, we need to know in which smoothness space $\mathcal{A}^s$ the solution lies. That is, we want to know the largest s such that the solution u is in $\mathcal{A}^s$. This is the role of regularity theorems for PDEs. The correct form for these theorems is a statement that says whenever the forcing data of the problem has certain properties then the solution lies in $\mathcal{A}^s$ for certain values of s.

The approximation and regularity theory tells us the optimal performance we can expect for our numerical method. However, they do not give us an actual numerical method since they usually use full information about u which is not available to us. So the third leg is to construct numerical methods which perform at this optimal performance. This entails not only the construction of the numerical method but also a rigorous analysis to establish its convergent rates.

We shall illustrate this trifecta by considering a very simple problem: the solution of Laplace's equation on a polygonal domain Ω in $\mathbb{R}^2$. Here we shall consider numerical methods of two distinct types. The first proceeds by specifying in advance the numerical method to be used. In the case of Finite Element Methods this means that a sequence (P_n) of triangulations of Ω are prescribed in advance. Typically, P_{n+1} is a uniform refinement of P_n. The solution u of the PDE is approximated by a piecewise polynomial subordinate to the partition P_n. When higher accuracy in the approximation is needed then n is increased.

There is a second class of numerical methods which does not set the full numerical scheme in advance but rather makes decisions on the run. In the case of Finite Element Methods, after an approximation to u has been made, the method examines the residual to decide what the new triangulation should be. Typically, some triangular cells are refined and others are not. These methods are called adaptive methods and we shall be interested in answering two question about their performance. The first is whether they provide any advantage over their non-adaptive counterparts? The second is how can we decide when an adaptive method is optimal?

1.4.1 Elliptic problems

We shall restrict our discussion to the Poisson problem

$$-\Delta u = f \quad \text{in} \ \ \Omega, \quad u = 0 \ \text{on} \ \partial\Omega, \tag{1.45}$$

where Ω is a polygonal domain in $\mathbb{R}^2$ and $\partial\Omega$ is its boundary. Our goal is to draw out the type of questions that should be asked when evaluating a numerical method for such elliptic problems.

In order to be able to work with the least smoothness requirements on approximations to u it is best to formulate (1.45), not in classical terms, but rather in its weak formulation. For this we introduce the Sobolev space $H_0^1(\Omega)$ of functions which vanish on the boundary $\partial\Omega$ of Ω and have weak derivatives of first order in $L_2(\Omega)$. The weak formulation of

(1.45) is to find $u \in H_0^1(\Omega)$ such that

$$a(u,w) = (f,w), \quad w \in H_0^1(\Omega), \tag{1.46}$$

where $a(y,w) := (\nabla y, \nabla w)$, with $(y,w) = (y,w)_\Omega := \int_\Omega ywdx$, is a quadratic form on $H_0^1(\Omega)$. Here f can be any distribution in $H^{-1}(\Omega)$. We use the notation

$$\|w\|^2 := a(w,w) = \|\nabla w\|^2_{L_2(\Omega)} \tag{1.47}$$

to denote the *energy norm* which is the natural norm in which to measure the performance of numerical methods. By Poincaré's inequality there exists a constant c_Ω, depending on Ω, such that for any $w \in H_0^1(\Omega)$,

$$c_\Omega \|w\|_{H^1(\Omega)} \leq \|w\| \leq \|w\|_{H^1(\Omega)}, \tag{1.48}$$

where $\|w\|^2_{H^1(\Omega)} = \|w\|^2_{L_2(\Omega)} + \|\nabla w\|^2_{L_2(\Omega)}$.

1.4.2 Newest vertex subdivision

The numerical methods we shall discuss are those which approximate the solution u of (1.46) by piecewise linear functions on triangular partitions of the polygonal domain Ω. The partitions are generated by refining triangles according to a fixed rule. While the following discussion applies to a variety of refinement rules we shall specify the method of subdivision to be *newest vertex bisection* since this will allow us a nice comparison when we discuss adaptive methods. The book of Verfürth [18] and the research article of Mitchell [15] describe this subdivision method and give some of its properties. The article [1] gives finer results on newest vertex bisection.

Let $P_0 = \{\Delta\}$ be an initial partition of Ω into a set of triangular cells. To each edge associated to this partition we assign a label of 0 or 1. We do this labelling in such a way that exactly one edge of any triangle has a label 0 and all other edges have a label 1 (such a labelling is always possible although it is nontrivial to prove; see [1]). For any triangle Δ of P_0, the vertex $v(\Delta)$ opposite the side of this triangle labelled as 0 is called its *newest vertex*. The edge in Δ opposite to $v(\Delta)$ will be denoted by $E(\Delta)$.

We have described how to assign to any triangle $\Delta \in P_0$ a newest vertex. Other triangles will arise by subdivision. We now describe the rule for subdividing triangles and how to assign a newest vertex to each triangle that arises by subdivision. If Δ is such a triangle and if $v(\Delta)$ is the newest vertex that has already been assigned then the subdivision of Δ consists of splitting Δ into two new triangles by inserting the diagonal that connects the newest vertex to the bisecting point of the edge $E(\Delta)$ opposite the newest vertex. Thus the cell produces two new cells and their newest vertex (assigned to each new triangular cell) is by definition the midpoint of $E(\Delta)$.

The partitions which arise when using newest vertex bisection satisfy a uniform minimal angle condition. This is established by showing that all triangles that arise in newest vertex bisection can be classified into a set of similarity classes depending only on the initial partition P_0 (see Mitchell).

We can represent newest vertex bisection by an infinite binary tree T_* (which we call the *master tree*). The master tree T_* consists of all triangular cells which can be obtained by a sequence of subdivisions. The roots of the master tree are the triangular cells in P_0. When a cell Δ is subdivided, it produces two new cells which are called the children of Δ and Δ is their parent. It is very important to note that, no matter how a cell arises in a subdivision process, its associated newest vertex is unique and only depends on the initial assignment of the newest vertices in P_0. This means that the children of Δ are uniquely determined and do not depend on how Δ arose in the subdivision process, i.e., it does not depend on the preceding sequence of subdivisions. The reason for this is that any subdivision only assigns newest vertices for the new triangular cells produced by the subdivision and does not alter any previous assignment. It follows that T_* is unique and does not depend at all on the order of subdivisions.

The *generation* of a triangular cell Δ is the number $g(\Delta)$ of ancestors it has in the master tree. Thus cells in P_0 have generation 0, their children have generation 1 and so on. The generation of a cell is also the number of subdivisions necessary to create this cell from its corresponding root cell in P_0.

There is a simple way to keep track of the newest vertices for triangular cells that arise in newest vertex bisection by giving a rule that labels any edges that arise from the subdivision process. There will be two main properties of this labelling. The first is that each triangular cell will have sides with labels $(i, i, i-1)$ for some positive integer i. The second is that the newest vertex for this cell will be the vertex opposite the side with lowest label. Certainly the edges in P_0 have such a labelling as we have just shown.

Suppose that a triangular cell $\Delta \in P_0$ has sides which have been labelled $(i, i, i-1)$ and the newest vertex for this cell is the one opposite the side labelled $i-1$. When this cell is subdivided (using newest vertex bisection) the side labelled $i-1$ is bisected and we label each of the two new sides $i+1$. We also label the *bisector* by $i+1$, i.e. the new edge connecting the newest vertex of Δ with the midpoint of the edge $E(\Delta)$ labelled by $i-1$. Thus each new triangle now has sides labelled $(i, i+1, i+1)$ with the newest vertex opposite the side with the lowest label. We note the important fact that if a cell has label $(i+1, i+1, i)$ then it is of generation i (i.e. it has been obtained from a cell in P_0 by i subdivisions). Therefore, specifying that the generation of the cell is i is the same as specifying that its label is $(i+1, i+1, i)$.

A *subtree* $T \subset T_*$ is a collection of triangular cells $\Delta \in T_*$ with the following two properties: (i) whenever $\Delta \in T$ then its sibling is also in the tree; (ii) when $\Delta \subset \Delta'$ are both in the tree then each triangular cell $\bar{\Delta} \in T_*$ with $\Delta \subset \bar{\Delta} \subset \Delta'$ is also in T. The roots of T are all the cells $\Delta \in T$ whose parents are not in T. We say that T is *proper* if it has the same roots as T_*, i.e., it contains all $\Delta \in P_0$. If $T \subset T_*$ is a finite subtree, we say $\Delta \in T$ is a *leaf* of T if T contains none of the children of Δ. We denote by $\mathcal{L}(T)$ the collection of all leaves of T. For a proper subtree T, we define $N(T)$ to be the number of subdivisions made to produce T.

Any partition $P = P_n$ which is obtained by the application of an adaptive procedure based on newest vertex bisection can be associated to a proper subtree $T = T(P)$ of T_* consisting of all triangular cells that were created during the algorithm, i.e. all of the cells in $P_0, \ldots, P_n$. The set of leaves $\mathcal{L}(T)$ form the final partition $P = P_n$.

We shall say that $T = T(P)$ and P are admissible if P has no hanging nodes. We denote the class of all proper trees by $\mathcal{T}$ and all admissible trees by $\mathcal{T}^a$. We also let $\mathcal{T}_n$ be the set of all proper trees T with $N(T) = n$ and by $\mathcal{T}_n^a$ the corresponding class of admissible trees from $\mathcal{T}_n$. We denote by $\mathcal{P}$ the class of all partitions P that can be generated by newest vertex bisection and by $\mathcal{P}^a$ the set of all admissible partitions. Similarly, $\mathcal{P}_n$ and $\mathcal{P}_n^a$ are the subclasses of those partitions that are obtained from P_0 by using n subdivisions. There is a precise identification between $\mathcal{P}_n$ and $\mathcal{T}_n$. Any $P \in \mathcal{P}_n$ can be given by a tree, i.e. $P = P(T)$ for some $T \in \mathcal{T}_n$. Conversely any $T \in \mathcal{T}_n$ determines a $P = P(T)$ in $\mathcal{P}_n$. The same can be said about admissible partitions and trees.

1.4.3 Standard finite element methods

As mentioned before, in this type of numerical approximation, a sequence of partitions $P_0, \ldots, P_n, \ldots$ is set in advance. In our case of newest vertex bisection, we start with P_0 and refine every triangular cell in P_0 (using newest vertex bisection) to obtain P_1, and continue in this way, refining every cell in a given P_n to obtain the next partition P_{n+1}. The partition P_n has the following properties. The triangular cells in P_n satisfy a minimal angle condition. Each angle is greater than a fixed positive constant c_0 which is independent of n (it depends only on the initial partition P_0). The partition P_n has no hanging nodes; this follows from the fact that every triangular cell is refined in moving from P_n to P_{n+1}. It follows that all triangular cells in P_n have comparable size. Their area is proportional to 2^{-n} and their diameter is proportional to $2^{-n/2}$.

Let $\mathcal{S}_n := \mathcal{S}_{P_n}$ be the space of continuous piecewise linear functions subordinate to the partition P_n which vanish on the boundary of Ω. These spaces are nested: $\mathcal{S}_n \subset \mathcal{S}_{n+1}$, $n = 0, 1, \ldots$. To numerically solve (1.45) we consider the Galerkin approximation to u from $\mathcal{S}_n$. It is determined by the system of equations

$$a(u_n, w) = (f, w), \quad w \in \mathcal{S}_n. \tag{1.49}$$

The solution u_n to this system minimizes the approximation in the energy norm:

$$\|u - u_n\| = \inf_{S \in \mathcal{S}_n} \|u - S\|. \tag{1.50}$$

In other words, given that we decided to measure the error in the energy norm and given that we have decided to use elements from $\mathcal{S}_n$ for the approximation, there is no question that we have found in u_n the best approximation to u that we can possibly obtain.

The role of approximation theory and regularity theory in this context is to explain what rate of approximation we can expect to obtain for a given right hand side f. The approximation theory says that we obtain an approximation rate

$$\|u - u_n\| = \mathcal{O}(2^{-ns/2}), \quad n = 1, 2, \ldots, \tag{1.51}$$

if and only if u is in the Besov space $B_\infty^{s+1}(L_2(\Omega))$. That is u should have s orders of smoothness more than the H^1 smoothness. Given the solution u to (1.45), we define $s_L = s_L(u)$ to be the maximum of all $s > 1$ such that $u \in B_\infty^s(L_2(\Omega))$. Here the subscript L indicates we are analyzing a linear approximation method; we will give a similar analysis for nonlinear (adaptive) methods.

Regularity theorems would tell us sufficient conditions on the right hand sides f for u to have a given Besov smoothness. In other words, they would give information on $s_L(u)$. For example, if we know that

$f \in L_2(\Omega)$ then u will at worst be in $B_\infty^{3/2}(L_2(\Omega))$. The worst behavior occurs for general Lipschitz domains. In our case of a polygonal domain the smoothness one can obtain depends on the angles corresponding to the boundary. In other words, for $f \in L_2(\Omega)$, the best we can say is that $s_L \geq 3/2$.

The point we wish to drive home here is that we know everything we could possibly want to know about our numerical method. We know it is the best method for approximating the solution by the approximation tools (piecewise linear functions) in the chosen metric (energy norm). We also know conditions on f sufficient to guarantee a given approximation rate and these sufficient conditions cannot be improved.

1.4.4 Adaptive finite element methods

We want to contrast the satisfying situation of the previous section with that for adaptive methods. Such methods do not set down the partitions P_n in advance but rather generate them in an adaptive fashion that depends very much on the solution u and the previous computations. We begin this part of the discussion by giving the form of a typical adaptive algorithm.

The starting point is, as before, the partition P_0. Given that the partition P_k has already been constructed, the algorithm computes the Galerkin solution u_k from $\mathcal{S}_{P_k}$. It examines $u_k = u_{P_k}$ and makes a decision whether or not to subdivide a given cell in P_k. In other words, the algorithm marks certain triangular cells $\Delta \in P_k$ for subdivision. We shall denote by $\mathcal{M}_k$ the collection of marked cells in P_k. These marked cells are subdivided using the newest vertex bisection. This process, however, creates *hanging nodes*. To remove the hanging nodes, a certain collection $\mathcal{M}'_k$ of additional cells are subdivided. The result after these two sets of subdivisions is the partition P_{k+1}. The partitions P_n, $n = 1, 2, \ldots$, have no hanging nodes and as noted earlier satisfy a minimal angle condition independent of n.

One of the keys to the success of such adaptive algorithms is the marking strategy, i.e. how to effectively choose which cells in P_k to subdivide. The idea is to choose only cells where the error $u - u_k$ is large (these would correspond to where the solution u is not smooth). Unfortunately, we do not know u (we are only seeing u through the computations u_k) so we do not know this local error. One uses in the place of the actual local error what are called *local error estimators*. The subject of local error estimators is a long and important one that we shall not enter into in this discussion. The reader may consult the the paper of Morin, Nochetto, and Siebert [16] for a discussion relevant to the present problem of Laplace's equation in two space dimensions. We mention only one important fact. The local error estimators allow one to compute accurate bounds for the global error $\|u - u_n\|$ from the knowledge of f and u_n. This is done through the residual $\Delta(u - u_n) = f - \Delta(u_n)$.

1.4.5 Judging the performance of adaptive methods

The main question we wish to address is how can we evaluate the performance of adaptive algorithms. Is there such a thing as an optimal or near optimal adaptive algorithm? This question is in the same spirit as the questions on image compression and A/D conversion. But there is a significant difference. In the other cases, the function to be encoded/approximated was completely known to us. In the PDE case, we only know u through our numerical approximations.

To begin the discussion, let us understand what approximants are available to us when we use adaptive methods. We have noted earlier that a partition P generated by an adaptive method can be associated to finite trees $T(P)$. As before, given any admissible partition P, we define $\mathcal{S}_P$ to be the space of piecewise linear functions which vanish on the boundary of Ω and are subordinate to P. For any function $w \in H_0^1(\Omega)$ and any such partition P, we define

$$E(w, \mathcal{S}_P) := \inf_{S \in \mathcal{S}_P} \|w - S\| = \|u - u_P\|, \qquad (1.52)$$

where u_P is the Galerkin approximation to u. This is the smallest error we can achieve by approximating u in the energy norm by the elements of $\mathcal{S}_P$.

How should we measure the complexity of an adaptively generated partition P. The most reasonable measure would be the number of computations used to create P. This turns out to be closely related to the number of subdivisions $n = N(T(P))$ used in the creation of P. As we shall see later, for certain specific adaptive algorithms, we can bound the number of computations used to create P and to compute u_P by Cn.

With these remarks in mind, we define

$$\sigma_n(u) := \inf_{P \in \mathcal{P}_n} E(w, \mathcal{S}_P) \qquad (1.53)$$

which is the best error we can possibly achieve in approximating w by using n subdivisions. It is unreasonable to expect any adaptive algorithm to perform exactly the same as $\sigma_n(w)$. However, we may expect the same *asymptotic* behavior. To quantify this, we introduce for any $s > 0$, the class $\mathcal{A}^s$ of functions $w \in H_0^1(\Omega)$ such that

$$\sigma_n(w) \leq Mn^{-s}, \quad n = 1, 2, \ldots. \tag{1.54}$$

The smallest M for which (1.54) is satisfied is the norm in $\mathcal{A}^s$:

$$\|w\|_{\mathcal{A}^s} := \sup_{n \geq 0} n^s \sigma_n(w). \tag{1.55}$$

We have a similar measure of approximation when we restrict ourselves to admissible partitions. Namely,

$$\sigma_n^a(w) := \inf_{P \in \mathcal{P}_n^a} E(w, \mathcal{S}_P)_{H^1(\Omega)} \tag{1.56}$$

now measures the best nonlinear approximation error obtained from admissible partitions and $\dot{\mathcal{A}}^s := \dot{\mathcal{A}}^s(H_0^1(\Omega))$ consists of all w which satisfy

$$\sigma_n^a(w) \leq Mn^{-s}, \quad n = 1, 2, \ldots. \tag{1.57}$$

The smallest M for which (1.57) holds serves to define the norm $\|w\|_{\dot{\mathcal{A}}^s}$. It is shown in [1] that the two spaces $\mathcal{A}^s$ are the same (see the discussion in §1.4.7); so we do not make a distinction between them in going further.

1.4.6 The role of regularity theorems

It is of interest to know which functions are in $\mathcal{A}^s$. In [2], it is shown that for a certain range of s (depending on the fact we are analyzing approximation by piecewise linear functions), we have

$$\mathcal{A}^{s/2} \approx B_p^s(L_p(\Omega)), \quad \frac{1}{p} = \frac{s-1}{2} + \frac{1}{2}. \tag{1.58}$$

The nebulous notation ($\approx$) used in this context means that there are certain embeddings between Besov spaces and the space $\mathcal{A}^s$. We do not want to get into the details (see [2]) but only work on this heuristic level. The spaces in (1.58) lie on the Sobolev embedding line for $H^1(\Omega)$ so they are very weak regularity conditions.

We can obtain a priori information on the possible performance of adaptive methods by knowing regularity theorems for the above scale of Besov spaces. For each u, let $s_{NL} := s_{NL}(u)$ be the maximum of all of the s such that $u \in B_p^s(L_p(\Omega))$, $\frac{1}{p} = \frac{s-1}{2} + \frac{1}{2}$. The larger the value of s_{NL}, the better performance we can hope for an adaptive algorithm. In particular, we would like to know if s_{NL} is larger than the corresponding value s_L for standard Finite Element Methods as discussed in §1.4.3. We emphasize that the above scale of Besov spaces that occur in describing the approximation rates for adaptive approximation are quite different from the scale of Besov spaces appearing in the linear case of standard Finite Element Methods. For a given $s > 0$ both require smoothness of order s but this smoothness is measured in different ways. In the standard case, the smoothness is relative to L_2 but in the adaptive case the smoothness is relative to an L_p which depends on s and gets smaller as s gets larger. The Besov space in the adaptive case is much larger than in the standard case and therefore more likely to contain u.

Let us take a simple example. We assume that the right side f is a function in $L_2(\Omega)$ and we ask what is the approximation error we can expect to receive if we invest n triangular cells to the approximation. As we have already described in §1.4.3, from this assumption on f we can always conclude that $f \in B^s_\infty(L_2(\Omega))$ for $s = 3/2$ and depending on the corners in the domain s could even be larger but never bigger than $s = 2$. Thus, $3/2 \leq s_L \leq 2$. This means that in general we cannot expect more than an approximation rate $\mathcal{O}(2^{\frac{-n(s_L-1)}{2}})$ when using standard finite element methods with 2^n triangular cells; see (1.51).

What is the situation for s_{NL}? To determine s_{NL} we need new regularity theorems that infer regularity in the new scale of Besov spaces $B^s_p(L_p(\Omega))$, with $\frac{1}{p} = \frac{s-1}{2} + \frac{1}{2}$. Regularity theorems of this type were proved in [8, 7]. For example, in the case of $f \in L_2(\Omega)$, it is shown that $s_{NL} \geq 2$ (see [8]). This means that for an investment of n triangular cells we should always obtain an error like $\mathcal{O}(n^{-1/2})$. This shows that potentially an adaptive method can perform much better than a standard Finite Element Method. If the standard method gives an error ϵ using a partition with n triangular cells, we might be able to obtain an error as small as ϵ^2 using the same number of cells in an adaptive method.

1.4.7 The performance of adaptive methods

The above analysis is all well and good but it provides us no information about specific numerical methods and does not even answer the question of whether such numerical methods converge. In fact, it was not until very recently that truly adaptive numerical methods were constructed and proven to converge (Dörfler [11], and Morin, Nochetto, Siebert [16]) and this convergence is only established for simple problems and special methods of subdivision. We are much more ambitious and would like to construct adaptive methods which perform at the optimal theoretical approximation rate.

Suppose we have an adaptive algorithm in hand that we have somehow constructed. Given an error tolerance $\epsilon > 0$, we run the adaptive algorithm until we achieve an error $< \epsilon$. We shall say that an algorithm is *near optimal* for s provided that whenever $u \in \mathcal{A}^s$ the number N of subdivisions used by the adaptive algorithm used to produce this error satisfies

$$N \leq C|u|_{\mathcal{A}^s}\epsilon^{-1/s}, \quad 0 < \epsilon < 1. \tag{1.59}$$

This means that the adaptive algorithm performs at the optimal rate on the class $\mathcal{A}^s$. The question of course is whether we can find an adaptive algorithm with this near optimal performance.

An adaptive algorithm with near optimal performance for solving (1.45) was constructed in [1]. Not only does this algorithm have optimality in terms of the number of subdivisions used but it is also shown that the number of computations $N(comp)$ needed to find u_{P_n} satisfies

$$N(comp) \leq C\epsilon^{-1/s}, \quad 0 < \epsilon < 1. \tag{1.60}$$

We shall not describe the algorithm of [1] in detail but instead give a brief overview of the algorithm and then pick up on details for the two core results that allow the proof of near optimality. The algorithm proceeds by setting error tolerances $\epsilon_k := 2^{-k}$ and generating partitions P_k, $k = 1, 2, \ldots$, in which the error is guaranteed to be $< \epsilon_k$. We stop at the first integer n where the error (which we can measure a posteriori) is $\leq \epsilon$. For sure, when

$$2^{-n} \leq \epsilon \tag{1.61}$$

this is the case. But the error bound ϵ may actually be attained for an earlier value of k.

The construction of P_k from P_{k-1} consists of several subiterations taking an admissible partition $P_{k-1,j}$ to a new admissible partition $P_{k-1,j+1}$. Here, $P_{k-1,0} = P_{k-1}$ and P_k is then found by coarsening a partition $P_{k-1,K}$ where K is a fixed integer. Each subiteration (sending $P_{k-1,j}$ to $P_{k-1,j+1}$) consists of the general strategy of marking cells, subdividing, further markings and subdividing to remove hanging nodes. The markings are made using the local error estimators found in [16] and marking enough cells so as to capture a fraction of the global error (this is known as bulk chasing). It was shown in [16] that such an iteration reduces the error in the energy norm. The role of K (which is a fixed integer) is to guarantee that after the coarsening step we have

$$\|u - u_{P_k}\| \leq 2^{-1}\|u - u_{P_{k-1}}\| \leq \epsilon_k = 2^{-k}. \tag{1.62}$$

Of course, (1.62) guarantees the algorithm converges but gives little information about convergence rates because we have said nothing about how many subdivisions were needed to construct P_k. However, a deeper analysis of the algorithm shows that it is near optimal, in the sense described above, for the range $0 < s < 1/2$ (we cannot expect to exceed this range because we are using piecewise linear functions to approximate in the $H^1(\Omega)$ norm).

The proof of near optimality rests on two fundamental results. The first of these concerns the number of subdivisions necessary to remove hanging nodes at a given iteration of the algorithm. It is necessary to have such an estimate to be sure that the number of triangular cells does not explode when the hanging nodes are removed. Suppose that we have a sequence of partitions $\tilde{P}_k$, $k = 1, \ldots, m$, where each $\tilde{P}_{k+1}$ is obtained from $\tilde{P}_k$ by marking a set $\tilde{\mathcal{M}}_k$, subdividing these cells, and then apply a further set of markings $\tilde{\mathcal{M}}'_k$ and subdivisions to remove hanging nodes. The fundamental result proved in [1] shows that

$$\#(\tilde{P}_m) \leq \#(\tilde{P}_0) + C_2(\#(\tilde{\mathcal{M}}_0) + \cdots + \#(\tilde{\mathcal{M}}_{m-1})). \tag{1.63}$$

This means that the number of markings used to remove hanging nodes is bounded by a multiple of the number of original markings. This is used in the adaptive finite element method to show, among other things, that

$$\#(P_k) \leq C\#(P_{k-1}) \tag{1.64}$$

with C an absolute constant.

The proof of (1.63) does not proceed, as one might expect, by an induction step which bound $\#(\tilde{\mathcal{M}}'_k)$ by $C\#(\tilde{\mathcal{M}}_k)$. Rather one needs to keep track of the entire history in the creation of new cells, tracing this history back to earlier cells that were subdivided; see [1] for details.

The second core result that is needed to keep control on the number of triangular cells and arrive at an estimate like (1.16) is that of coarsening. Without this step, we would not arrive (1.59). The coarsening step proceeds as follows. After constructing the set $P_{k,K}$, we compute a Galerkin approximation $v_k := u_{P_{k,K}}$ to u from $\mathcal{S}_{P_{k,K}}$ which satisfies

$$\|u - v_k\| \leq \gamma \epsilon_{k+1} \tag{1.65}$$

with γ a small fixed positive constant. The choice of γ influences the choice of K above. The function v_k is known to us and the coarsening algorithm uses information about this function to create a smaller partition $P_{k+1} \subset P_{k,K}$ such that v_k can be approximated by an element of $\mathcal{S}_{P_{k+1}}$ to accuracy $\alpha \epsilon_k$ where again $0 < \alpha < 1$ is a fixed constant that is chosen appropriate to the discussion that follows. The property gained by the coarsening step is that P_{k+1} satisfies

$$\#(P_{k+1}) \leq C_0 |u|_{\mathcal{A}^s} \epsilon_k^{-1/s} \tag{1.66}$$

with C_0 an absolute constant. It is then shown that the Galerkin approximation $u_{k+1} = u_{P_{k+1}}$ satisfies

$$\|u - u_{k+1}\| \leq \frac{\epsilon_k}{2} = \epsilon_{k+1}. \tag{1.67}$$

This estimate together with the control on the size of the partition P_{k+1} given in (1.66) combine to show that the algorithm is near optimal.

It is vital in the coarsening algorithm to keep a control on the number of computations needed to find P_{k+1}. This is in fact the deeper aspect of the coarsening algorithm. Normally to find a near optimal tree approximation with n nodes, one might expect to have to do exponential in n computations since there are that many trees with n nodes. It is shown in [3] that the number of computations can be limited to $\mathcal{O}(n)$ by using a clever adaptive strategy.

References

[1] Binev, P., Dahmen, W. and DeVore, R. (2002). Adaptive finite element methods with convergence rates, *IMI Preprint Series* **12** (University of South Carolina).
[2] Binev, P., Dahmen, W., DeVore, R. and Petrushev, P. (2002). Approximation Classes for Adaptive Methods, *Serdica Math. J.* **28**, 391–416.
[3] Binev, P. and DeVore, R. (2002). Fast computation in tree approximation, *IMI Preprint Series* **11** (University of South Carolina).
[4] Calderbank, R. and Daubechies, I. (2002). The Pros and Cons of Democracy *IEEE Transactions in Information Theory* **48**, 1721–1725.
[5] Cohen, A., Dahmen, W., Daubechies, I. and DeVore, R. (2001). Tree approximation and encoding, *ACHA* **11**, 192–226.
[6] Cohen, A., Daubechies, I., Guleryuz, O. and Orchard, M. (2002). On the importance of combining non-linear approximation with coding strategies, *IEEE Transactions in Information Theory* **48**, 1895–1921.
[7] Dahlke, S. (1999). Besov regularity for elliptic boundary value problems on polygonal domains, *Appl. Math. Lett.* **12**, 31–36.
[8] Dahlke, S. and DeVore, R. (1997). Besov regularity for elliptic boundary value problems, *Comm. Partial Differential Equations* **22**, 1–16.

[9] Daubechies, I. and DeVore, R. Reconstructing a bandlimited function from very coarsely quantized data: A family of stable sigma-delta modulators of arbitrary order, *Annals of Mathematics*, to appear.
[10] Daubechies, I., DeVore, R., Gunturk, C.S. and Vaishampayan, V. Exponential Precision in A/D Conversion with an Imperfect Quantizer, preprint.
[11] Dörfler, W. (1996). A convergent adaptive algorithm for Poisson's equation, *SIAM J. Numer. Anal.* **33**, 1106–1124.
[12] Güntürk, S. One-bit Sigma-Delta quantization with exponential accuracy, preprint.
[13] Kolmogorov, A.N. and Tikhomirov, V.M. (1959). ϵ-entropy and ϵ-capacity of sets in function spaces, *Usephi* **14**, 3–86. (Also in *AMS Translations, Series 2* **17** (1961), 277–364).
[14] Meyer, Y. (2001). Oscillating pattern in image processing and in some nonlinear evolution equations, *The Fifteenth Dean Jaqueline B. Lewis Memorial Lectures*.
[15] Mitchell, W.F. (1989). A comparison of adaptive refinement techniques for elliptic problems, *ACM Transaction on Math. Software* **15**, 326–347.
[16] Morin, P., Nochetto, R. and Siebert, K. (2000). Data Oscillation and Convergence of Adaptive FEM, *SIAM J. Numer. Anal.* **38**, 466–488.
[17] Mumford, D. and Shah, J. (1985). Boundary detection by minimizing functionals, *Proc. IEEE Conf. Comp. Vis Pattern Recognition*.
[18] Verfürth, R. (1996). *A Review of A Posteriori Error Estimation and Adaptive Mesh-Refinement Techniques* (Wiley-Teubner, Chichester).

2

Jacobi Sets of Multiple Morse Functions

Herbert Edelsbrunner

Department of Computer Science and Mathematics
Duke University, Durham, and
Raindrop Geomagic
Research Triangle Park
North Carolina

John Harer

Department of Computer Science and Mathematics
Duke University, Durham
North Carolina

Abstract

The Jacobi set of two Morse functions defined on a common d-manifold is the set of critical points of the restrictions of one function to the level sets of the other function. Equivalently, it is the set of points where the gradients of the functions are parallel. For a generic pair of Morse functions, the Jacobi set is a smoothly embedded 1-manifold. We give a polynomial-time algorithm that computes the piecewise linear analog of the Jacobi set for functions specified at the vertices of a triangulation, and we generalize all results to more than two but at most d Morse functions.

2.1 Introduction

This paper is a mathematical and algorithmic study of multiple Morse functions, and in particular of their Jacobi sets. As we will see, this set is related to the Lagrange multiplier method in multi-variable calculus of which our algorithm may be viewed as a discrete analog.

Motivation. Natural phenomena are frequently modeled using continuous functions, and having two or more such functions defined on the same domain is a fairly common scenario in the sciences. Consider for

---

[0] Research of the first author is partially supported by NSF under grants EIA-99-72879 and CCR-00-86013. Research of the second author is partially supported by NSF under grant DMS-01-07621.

example oceanography where researchers study the distribution of various attributes of water, with the goal to shed light on the ocean dynamics and gain insight into global climate changes [4]. One such attribute is temperature, another is salinity, an important indicator of water density. The temperature distribution is often studied within a layer of constant salinity, because water tends to mix along but not between these layers. Mathematically, we may think of temperature and salinity as two continuous functions on a common portion of three-dimensional space. A layer is determined by a level surface of the salinity function, and we are interested in the temperature function restricted to that surface. This is a continuous function on a two-dimensional domain, whose critical points are generically minima, saddles, and maxima. In this paper, we study the paths these critical points take when the salinity value varies. As it turns out, these paths are also the paths the critical points of the salinity function take if we restrict it to the level surfaces of the temperature function. More generally, we study the relationship between continuous functions defined on a common manifold by analyzing the critical points within level set restrictions.

Sometimes it is useful to make up auxiliary functions to study the properties of given ones. Consider for example a function that varies with time, such as the gravitational potential generated by the sun, planets, and moons in our solar system [18]. At the critical points of that potential, the gravitational forces are at an equilibrium. The planets and moons move relative to each other and the sun, which implies that the critical points move, appear, and disappear. To study such a time-varying function, we introduce another, whose value at any point in space-time is the time. The paths of the critical points of the gravitational potential are then the Jacobi set of the two functions defined on a common portion of space-time.

Results. The main object of study in this paper is the Jacobi set $\mathbb{J} = \mathbb{J}(f_0, f_1, \ldots, f_k)$ of $k+1 \leq d$ Morse functions on a common d-manifold. By definition, this is the set of critical points of f_0 restricted to the intersection of the level sets of f_1 to f_k. We observe that $\mathbb{J}$ is symmetric in the $k+1$ functions because it is the set of points at which the $k+1$ gradient vectors are linearly dependent. In the simplest non-trivial case, we have two Morse functions on a common 2-manifold. In this case, the Jacobi set $\mathbb{J} = \mathbb{J}(f,g) = \mathbb{J}(g,f)$ is generically a collection of pairwise disjoint smooth curves that are free of any self-intersections. Fig. 2.1 illustrates the concept for two Morse functions on the two-dimensional

2. Jacobi Sets

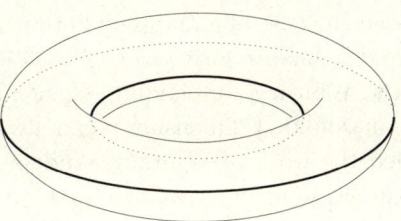

Fig. 2.1. The two partially bold and partially dotted longitudinal circles form the Jacobi set of $f, g : \mathbb{M} \to \mathbb{R}$, where $\mathbb{M}$ is the torus and f and g map a point $x \in \mathbb{M}$ to the Cartesian coordinates of its orthogonal projection on a plane parallel to the longitudes.

torus. The Jacobi set of a generic collection of $k + 1$ Morse functions is a submanifold of dimension k, provided $d > 2k - 2$. The first time this inequality fails is for $d = k + 1 = 4$. In this case, the Jacobi set is a 3-manifold except at a discrete set of points.

We describe an algorithm that computes an approximation of the Jacobi set for $k+1$ piecewise linear functions that are interpolated from values given at the vertices. In the absence of smoothness, it simulates genericity and differentiability and computes $\mathbb{J}$ as a subcomplex of the triangulation. The algorithm is combinatorial (as opposed to numerical) in nature and reduces the computations to testing the criticality of the k-simplices in the triangulation. Whether or not a simplex is considered critical depends on its local topology. By using Betti numbers to express that topology, we get an algorithm that works for triangulated manifolds and runs in time that is polynomial in the number of simplices. Assuming the links have sizes bounded from above by a constant, the running time is proportional to the number of simplices.

Related prior work. The work in this paper fits within the general area of singularities of smooth mappings, which was pioneered by Hassler Whitney about half a century ago; see e.g. [19]. In this body of work, the *fold* of a mapping from a d-dimensional manifold $\mathbb{M}^d$ to a $(k+1)$-dimensional manifold $\mathbb{N}^{k+1}$ is the image of the set of points at which the matrix of partial derivatives has less than full rank. In this paper, we consider $(k+1)$-tuple of Morse functions, which are special smooth mappings in which the range is the $(k+1)$-dimensional Euclidean space, $\mathbb{R}^{k+1}$. The fold of such a special mapping is the image of the Jacobi set. Our restriction to Euclidean space is deliberate as it furnishes the framework needed for our algorithm.

Whitney considered the case of a mapping between surfaces and studied mappings where d is small relative to k. A classic theorem in differential topology is the Whitney Embedding Theorem which states that a closed, orientable manifold of dimension n can always be embedded in $\mathbb{R}^{2n}$. Thom extended the work of Whitney, studying the spaces of jets (two functions have the same k-jet at p if their partial derivatives of order k or less are equal). The Thom Transversality Theorem, together with the Thom-Boardman stratification of C^∞ functions on $\mathbb{M}$, give characterizations of the singularities of generic functions [11]. Mather studied singularities from a more algebro-geometric point of view and proved the equivalence of several definitions of stability for a map (a concept not used in this work). In particular he provided the appropriate framework to reconstruct a map from its restrictions to the strata of the Thom-Boardman stratification.

In a completely different context, Nicola Wolpert used Jacobi curves to develop exact and efficient algorithms for intersecting quadratic surfaces in $\mathbb{R}^3$ [20]. She does so by reducing the problem to computing the arrangement of the intersection curves projected to $\mathbb{R}^2$. Any such curve can be written as the zero set of a smooth function $\mathbb{R}^2 \to \mathbb{R}$, and any pair defines another curve, namely the Jacobi set of the two functions. Wolpert refers to them as *Jacobi curves* and uses them to resolve tangential and almost tangential intersections. To be consistent with her terminology, we decided to modify that name and to refer to our more general concept as Jacobi sets.

Outline. §2.2 reviews background material from differential and combinatorial topology. §2.3 introduces the Jacobi set of two Morse functions. §2.4 describes an algorithm that computes this set for piecewise linear data. §2.5 generalizes the results to three or more Morse functions. §2.6 discusses a small selection of applications. §2.7 concludes the paper.

2.2 Background

This paper contains results for smooth and for piecewise linear functions. We need background from Morse theory [15, 16] for smooth functions and from combinatorial and algebraic topology [1, 17] for designing an algorithm that works on piecewise linear data.

Morse functions. Let $\mathbb{M}$ be a smooth and compact d-manifold without boundary. The differential of a smooth map $f : \mathbb{M} \to \mathbb{R}$ at a point x of the manifold is a linear map $\mathrm{d}f_x : \mathrm{TM}_x \to \mathbb{R}$ mapping the tangent space at x to $\mathbb{R}$. A point $x \in \mathbb{M}$ is *critical* if $\mathrm{d}f_x$ is the zero map; otherwise, it is *regular*. Let $\langle , \rangle$ be a Riemannian metric, i.e. an inner product in the tangent spaces that varies smoothly on $\mathbb{M}$. Since each vector in TM_x is the tangent vector to a curve c in $\mathbb{M}$ through x, the *gradient* ∇f can be defined by the formula $\langle \partial c / \partial t, \nabla f \rangle = \partial(f \circ c)/\partial t$, for every c. It is always possible to choose coordinates x_i so that the tangent vectors $\frac{\partial}{\partial x_i}(x)$ are orthonormal with respect to the Riemannian metric. For such coordinates, the gradient is given by the familiar formula

$$\nabla f(x) = \left[\frac{\partial f}{\partial x_1}(x), \frac{\partial f}{\partial x_2}(x), \ldots, \frac{\partial f}{\partial x_d}(x) \right].$$

We compute in local coordinates the *Hessian* of f:

$$H_f(x) = \begin{bmatrix} \frac{\partial^2 f}{\partial x_1^2}(x) & \cdots & \frac{\partial^2 f}{\partial x_d \partial x_1}(x) \\ \vdots & \ddots & \vdots \\ \frac{\partial^2 f}{\partial x_1 \partial x_d}(x) & \cdots & \frac{\partial^2 f}{\partial x_d^2}(x) \end{bmatrix},$$

which is a symmetric bi-linear form on the tangent space TM_x. A critical point p is *non-degenerate* if the Hessian is non-singular at p. The Morse Lemma states that near a non-degenerate critical point p, it is possible to choose local coordinates so that the function takes the form

$$f(x_1, \ldots, x_d) = f(p) \pm x_1^2 \pm \ldots \pm x_d^2.$$

The number of minus signs is the *index* of p; it equals the number of negative eigenvalues of $H_f(p)$. The existence of these local coordinates implies that non-degenerate critical points are isolated. The function f is a called a *Morse function* if

(i) all its critical points are non-degenerate, and
(ii) $f(p) \neq f(q)$ for all critical points $p \neq q$.

Transversality and stratification. Let $F : \mathbb{X} \to \mathbb{Y}$ be a smooth map between two manifolds, and let $\mathbb{U} \subseteq \mathbb{Y}$ be a smooth submanifold. The map F is *transversal* to $\mathbb{U}$ if for every $y \in \mathbb{U}$ and every $x \in F^{-1}(y)$, we have $\mathrm{d}F_x(\mathrm{T}\mathbb{X}_x) + \mathrm{T}\mathbb{U}_y = \mathrm{T}\mathbb{Y}_y$. In words, the basis vectors of the image of the tangent space of $\mathbb{X}$ at x under the derivative together with the basis vectors of the tangent space of $\mathbb{U}$ at $y = F(x)$ span the tangent space of $\mathbb{Y}$ at y. The Transversality Theorem of differential topology asserts that if F is transversal to $\mathbb{U}$ then $F^{-1}(\mathbb{U})$ is a smooth submanifold of $\mathbb{X}$ and the co-dimension of $F^{-1}(\mathbb{U})$ in $\mathbb{X}$ is the same as that of $\mathbb{U}$ in $\mathbb{Y}$ [12].

A continuous family connecting two Morse functions necessarily goes through transitions at which the function violates conditions (i) and (ii) of a Morse function. We are interested in the minimum number of violations that cannot be avoided. For this purpose, consider the infinite-dimensional Hilbert space $C^\infty(\mathbb{M})$ of smooth functions $\mathbb{M} \to \mathbb{R}$. There is a stratification $C^\infty(\mathbb{M}) = C_0 \supset C_1 \supset C_2$, in which $C_0 - C_1$ is the set of Morse functions, $C_1 - C_2$ is the set of functions that violate either condition (i) or (ii) exactly once, and C_2 is the set of remaining functions. The set $C_0 - C_1$ is a submanifold of co-dimension 0, $C_1 - C_2$ is a submanifold of co-dimension 1, and C_2 has co-dimension 2. As an illustration of how we use the stratification, take two Morse functions f and g and consider the 1-parameter family of functions $h_\lambda = f + \lambda g : \mathbb{M} \to \mathbb{R}$. We can perturb one of the functions such that

- f and $f + g$ are both Morse,
- all h_λ belong to $C_0 - C_2$, and
- the 1-parameter family is transversal to $C_1 - C_2$.

If follows that h_λ belongs to $C_1 - C_2$ for only a discrete collection of values for λ. We use this later to prove the transversality of certain maps, which will then put us into the position to apply the Transversality Theorem.

Triangulations. A *k-simplex* is the convex hull of $k+1$ affinely independent points. Given two simplices σ and τ, we write $\tau \leq \sigma$ if τ is a face of σ. A *simplicial complex* is a finite collection K of simplices that is closed under the face relation such that the intersection of any two simplices is either empty or a face of both. A *subcomplex* of K is a subset that itself is a simplicial complex. The *closure* of a subset $L \subseteq K$ is the smallest subcomplex $\operatorname{Cl} L \subseteq K$ that contains L. The *star* of a simplex $\tau \in K$ is the collection of simplices that contain τ, and the *link* is the collection of faces of simplices in the closure of the star that are disjoint from τ:

$$\operatorname{St} \tau = \{\sigma \in K \mid \tau \leq \sigma\},$$
$$\operatorname{Lk} \tau = \{\upsilon \in \operatorname{Cl} \operatorname{St} \tau \mid \upsilon \cap \tau = \emptyset\}.$$

The *vertex set* of K is denoted by $\operatorname{Vert} K$. The *underlying space* is the union of simplices: $|K| = \bigcup_{\sigma \in K} \sigma$. The *interior* of a simplex σ is the set of points that belong to σ but not to any proper face of σ. Note that each point of $|K|$ belongs to the interior of exactly one simplex in K. We specify a piecewise linear continuous function by its values at the vertices. To describe this construction, let $\varphi : \operatorname{Vert} K \to \mathbb{R}$ be a function defined on the vertex set. Each point $x \in |K|$ has barycentric coordinates $\alpha_u = \alpha_u(x) \in \mathbb{R}$ such that $\sum \alpha_u u = x$, $\sum \alpha_u = 1$, and $\alpha_u = 0$ unless u is a vertex of the simplex whose interior contains x. The *linear extension* of φ is the function $f : |K| \to \mathbb{R}$ defined by $f(x) = \sum_u \alpha_u(x) \varphi(x)$. It is continuous because the maps α_u are continuous.

The simplicial complex K is a *triangulation* of a manifold $\mathbb{M}$ if there is a homeomorphism $|K| \to \mathbb{M}$. In this case, the link of every k-simplex triangulates a sphere of dimension $d - k - 1$. We will extend the concept of a critical point from smooth to piecewise linear functions using subcomplexes of links. Assuming $f(u) \neq f(v)$ for any two vertices $u \neq v$ in K, the *lower star* of u consists of all simplices in the star that have u as their highest vertex, and the *lower link* is the portion of the link that bounds the lower star. Note that both the link and the lower link are always complexes.

Homology groups and Betti numbers. In the piecewise linear case, we use the topology of the lower link to define whether or not we consider a vertex critical, and we express that topology using ranks of homology groups for $\mathbb{Z}_2$ coefficients. To explain this, let K be a simplicial complex. A k-*chain* is a subset of the k-simplices, and *adding* two k-chains means taking their symmetric difference. The set of k-chains together with addition forms the group of k-chains, denoted as C_k. The *boundary* of a k-simplex is the set of its $(k-1)$-simplices. This defines a *boundary homomorphism* $\partial_k : \mathsf{C}_k \to \mathsf{C}_{k-1}$ obtained by mapping a k-chain to the sum of boundaries of its k-simplices. Since we aim at reduced homology groups, we extend the list of non-trivial chain groups by adding $\mathsf{C}_{-1} = \mathbb{Z}_2$ and defining the boundary of a vertex as $\partial_0(u) = 1$. The boundary homomorphisms connect the chain groups as illustrated in Fig. 2.2. A k-*cycle* is a k-chain with zero boundary, and a k-*boundary* is the boundary of

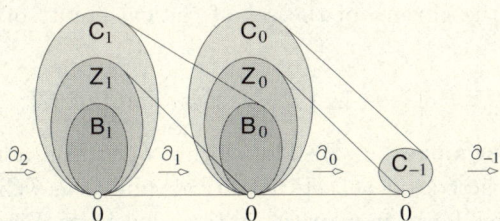

Fig. 2.2. The chain complex formed by the chain groups and the connecting boundary homomorphisms.

a $(k+1)$-chain. The corresponding groups are nested subgroups of the group of k-chains: $\mathsf{B}_k \leq \mathsf{Z}_k \leq \mathsf{C}_k$. The *k-th reduced homology group* is the quotient defined by the k-cycles and the k-boundaries: $\tilde{\mathsf{H}}_k = \mathsf{Z}_k/\mathsf{B}_k$. The *k-th reduced Betti number* is the rank of the k-th reduced homology group: $\tilde{\beta}_k = \operatorname{rank} \tilde{\mathsf{H}}_k$. Since we add modulo 2, all groups are finite and $\tilde{\beta}_k$ is the binary logarithm of the size of $\tilde{\mathsf{H}}_k$. The reduced homology groups differ from the more common non-reduced versions only in dimensions 0 and -1. Specifically, $\tilde{\beta}_0$ is one less than the number of components, unless the complex is empty, in which case $\tilde{\beta}_0 = 0$, and $\tilde{\beta}_{-1} = 0$ unless the complex is empty, in which case $\tilde{\beta}_{-1} = 1$.

The main reason for preferring reduced over non-reduced homology groups is their simpler correspondence to spheres. The $(d-1)$-*sphere* is the set of points at unit distance from the origin of the d-dimensional Euclidean space: $\mathbb{S}^{d-1} = \{x \in \mathbb{R}^d \mid \|x\| = 1\}$. To triangulate $\mathbb{S}^{d-1}$ we may take the set of proper faces of a d-simplex. Note that $\mathbb{S}^{-1}$ is the empty set, which is triangulated by the empty complex. The $(d-1)$-sphere has only one non-zero reduced Betti number, namely $\tilde{\beta}_{d-1} = 1$, and this is true for all $d \geq 0$. In contrast, the reduced Betti numbers of a point all vanish.

2.3 Jacobi sets of two functions

In this section, we consider the Jacobi set of a pair of Morse functions defined on the same manifold.

Definition of Jacobi sets. Let $\mathbb{M}$ be a smooth d-manifold and choose a Riemannian metric so that we can define gradients. We assume $d \geq 2$ and consider two generic Morse functions, $f, g : \mathbb{M} \to \mathbb{R}$. We are interested in the restrictions of f to the level sets of g. For a regular value $t \in \mathbb{R}$, the level set $\mathbb{M}_t = g^{-1}(t)$ is a smooth $(d-1)$-manifold, and the restriction of f to this level set is a Morse function $f_t : \mathbb{M}_t \to \mathbb{R}$. The *Jacobi set* $\mathbb{J} = \mathbb{J}(f, g)$ is the closure of the set of critical points of such level set restrictions:

$$\mathbb{J} = \operatorname{cl}\{x \in \mathbb{M} \mid x \text{ is critical point of } f_t\}, \qquad (2.1)$$

for some regular value $t \in \mathbb{R}$. The closure operation adds the critical points of f restricted to level sets at critical values t as well as the critical points of g, which form singularities in these level sets. Fig. 2.3 illustrates the definition by showing $\mathbb{J}$ for two smooth functions defined on the two-dimensional plane. Think of the picture as a cone-like mountain

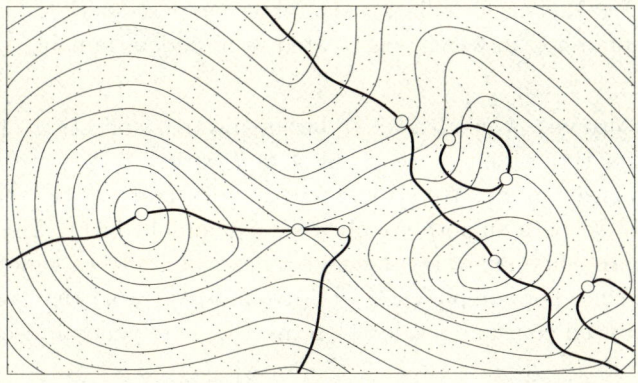

Fig. 2.3. The functions f and g are represented by their dotted and solid level curves. The Jacobi set is drawn in bold solid lines. The birth-death points and the critical points of the two functions are marked.

indicated by the (dotted) level curves of f, and imagine an animation during which the (solid) level curves of g glide over that mountain. For example, on the left we see a circle expanding outwards on a slope, and we observe a maximum moving uphill and a minimum moving downhill from the starting point, which is a minimum of g.

Consider the gradient of the two functions at a point x: $\nabla f(x), \nabla g(x) \in \mathbb{R}^d$. Let $t = g(x)$. Then the gradient of f_t at $x \in \mathbb{M}_t = g^{-1}(t)$ is the projection of $\nabla f(x)$ onto the tangent space of $\mathbb{M}_t$ at x. It follows that $\nabla f_t(x) = 0$ iff the two gradient vectors are parallel. In other words,

$$\mathbb{J} = \{x \in \mathbb{M} \mid \nabla f(x) + \lambda \nabla g(x) = 0 \text{ or } \\ \lambda \nabla f(x) + \nabla g(x) = 0\}, \quad (2.2)$$

for some $\lambda \in \mathbb{R}$. The first relation in (2.2) misses the cases in which $\nabla g(x) = 0$ and $\nabla f(x) \neq 0$. The only reason for the second relation is to capture these cases. The description of $\mathbb{J}$ in (2.2) is symmetric in f and g: $\mathbb{J}(f,g) = \mathbb{J}(g,f)$. We therefore call the points of $\mathbb{J}$ *simultaneous critical points* of f and g. Note that $\nabla f + \lambda \nabla g$ is the gradient of the function $f + \lambda g$. Equation (2.2) thus implies yet another characterization of the Jacobi set:

$$\mathbb{J} = \{x \in \mathbb{M} \mid x \text{ is critical point of} \\ f + \lambda g \text{ or of } \lambda f + g\}, \quad (2.3)$$

for some $\lambda \in \mathbb{R}$. This formulation should be compared with the Lagrange multiplier method in which λ is treated as an indeterminant.

Critical curves. Generically, f_t has a discrete collection of critical points, and these points sweep out $\mathbb{J}$ as t varies. It follows that $\mathbb{J}$ is a one-dimensional set. We strengthen this observation and prove that the Jacobi set is a smoothly embedded 1-manifold in $\mathbb{M}$. This follows form the stratification of the Jacobi set described in [11], which we will sketch in §2.5. For completeness, we give a direct proof that avoids the more advanced concepts needed to prove the more general result.

SMOOTH EMBEDDING THEOREM. *Generically, the Jacobi set of two Morse functions $f, g : \mathbb{M} \to \mathbb{R}$ is a smoothly embedded 1-manifold in $\mathbb{M}$.*

Proof Assume $\mathbb{M}$ is a d-manifold, for $d \geq 2$, and consider the functions $F, G : \mathbb{M} \times \mathbb{R} \to \mathbb{R}^d$ that map a point $x \in \mathbb{M}$ and a parameter $\lambda \in \mathbb{R}$ to the gradients of $f + \lambda g$ and $\lambda f + g$:

$$F(x, \lambda) = \nabla f(x) + \lambda \nabla g(x),$$
$$G(x, \lambda) = \lambda \nabla f(x) + \nabla g(x).$$

By (2.2), a point x belongs to $\mathbb{J}$ iff there is a $\lambda \in \mathbb{R}$ such that $F(x, \lambda) = 0$ or $G(x, \lambda) = 0$. Letting $\Gamma_F = F^{-1}(0)$ and $\Gamma_G = G^{-1}(0)$, we get $\mathbb{J}$ by projecting onto $\mathbb{M}$:

$$\begin{array}{ccccc} \Gamma_F \subseteq \mathbb{M} \times \mathbb{R} & & & \Gamma_G \subseteq \mathbb{M} \times \mathbb{R} \\ \downarrow \quad \downarrow & & \text{and} & \downarrow \quad \downarrow \\ \mathbb{J}_F \subseteq \mathbb{M} & & & \mathbb{J}_G \subseteq \mathbb{M}. \end{array}$$

We have $\mathbb{J} = \mathbb{J}_F \cup \mathbb{J}_G$, where $\mathbb{J}_F$ is $\mathbb{J}$ minus the discrete set of points where $\nabla g(p) = 0$, and $\mathbb{J}_G$ is $\mathbb{J}$ minus the points where $\nabla f(p) = 0$. We prove that F is transversal to 0 or, equivalently, that for every $(p, \lambda) \in F^{-1}(0)$, the derivative of F at (p, λ) has rank d. We compute $dF_{(p,\lambda)} = [H_{f+\lambda g}(p), \nabla g(p)]$, where the Hessian is a d-by-d matrix. As mentioned in §2.2, we can assume that all functions $h_\lambda = f + \lambda g$ are in $C_0 - C_1$, except for a discrete number, which are in $C_1 - C_2$. The former have only non-degenerate critical points, so the Hessian itself already has rank d. Let λ_0 be a value for which h_{λ_0} is not Morse. If there are two critical points sharing the same function value, then the Hessian is still invertible and there is nothing else to show. Otherwise, there is a single birth-death

2. Jacobi Sets

point p_0, and we write $c_0 = h_{\lambda_0}(p_0)$. There exist local coordinates such that $p_0 = (0, 0, \ldots, 0)$, $\lambda_0 = 0$, and

$$h_\lambda(p) = c_0 + x_1^3 - \lambda x_1 \pm x_2^2 \pm \ldots \pm x_d^2$$
$$= \lambda(-x_1) + (c_0 + x_1^3 \pm x_2^2 \pm \ldots \pm x_d^2)$$

in a neighborhood of (p_0, λ_0). Note that this implies $g(p) = -x_1$. We can write the Hessian and the gradient explicitly and get

$$dF_{(p_0, \lambda_0)} = \begin{bmatrix} 0 & 0 & \ldots & 0 & -1 \\ 0 & \pm 2 & \ldots & 0 & 0 \\ \vdots & \vdots & \ddots & \vdots & \vdots \\ 0 & 0 & \ldots & \pm 2 & 0 \end{bmatrix}.$$

This matrix has rank d. Since $0 \in \mathbb{R}^d$ has co-dimension d, the Transversality Theorem now implies that $\Gamma_F = F^{-1}(0)$ is a smooth 1-manifold in $\mathbb{M} \times \mathbb{R}$. We still need to prove that the projection of this 1-manifold is smoothly embedded in $\mathbb{M}$. Let $\pi : \Gamma_F \to \mathbb{J}$ be defined by $\pi(p, \lambda) = p$. We show that π is one-to-one and $d\pi_{(p, \lambda)} \neq 0$ for all $(p, \lambda) \in \Gamma_F$. If $\pi(p, \lambda) = \pi(p', \lambda')$ then $p = p'$ and $\nabla f(p) + \lambda \nabla g(p) = \nabla f(p) + \lambda' \nabla g(p) = 0$. Since $\nabla g(p) \neq 0$ on Γ_F, $\lambda = \lambda'$ so the projection is injective. To prove that the derivative $d\pi$ of π is non-zero, note that the tangent line to Γ_F at (p, λ) is the kernel of $dF_{(p, \lambda)} = [H_{f + \lambda g}(p), \nabla g(p)]$. If $d\pi$ is zero at (p, λ), then this tangent line must be vertical, spanned by the vector $v = (0, \ldots, 0, 1)$. But since $\nabla g(p) \neq 0$, v cannot be in the kernel of $dF(p, \lambda)$.

By symmetry, everything we proved for F also holds for G. This concludes our proof that generically, $\mathbb{J}$ is a smoothly embedded 1-manifold in $\mathbb{M}$. □

2.4 Algorithm

In this section, we describe an algorithm that computes an approximation of the Jacobi set of two Morse functions from approximations of these functions. We begin by laying out our general philosophy and follow up by describing the details of the algorithm.

General approach. In applications, we never have smooth functions but usually non-smooth functions that approximate smooth functions. We choose not to think of the non-smooth functions as approximations of smooth functions. Quite the opposite, we think of the non-smooth functions as being approximated by smooth functions. The relatively simple smooth case then becomes the guiding intuition in designing the algorithm that constructs what one may call the Jacobi set of the non-smooth functions. To be more specific, let K be a triangulation of a d-manifold $\mathbb{M}$ and let φ, ψ : Vert $K \to \mathbb{R}$ be two functions defined at the vertices. We obtain $f, g : |K| \to \mathbb{R}$ as piecewise linear extensions of φ and ψ. We imagine that both piecewise linear functions are limits of series of smooth functions: $\lim_{n\to\infty} f_n = f$ and $\lim_{n\to\infty} g_n = g$. For each n, the Jacobi set $\mathbb{J}_n = \mathbb{J}(f_n, g_n)$ is perfectly well defined, and we aim at constructing the Jacobi set of f and g as the limit of the $\mathbb{J}_n$. Along the way, we will take some liberties in resolving the ambiguities in what this means exactly. As a guiding principle, we resolve ambiguities in a way consistent with the corresponding smooth concepts that arise in the imagined smooth approximations of f and g. We call this principle the *simulation of differentiability* and combine it with the simulation of simplicity [9] to put ourselves within the realm of generic smooth functions.

After constructing the Jacobi set of f and g, we may need to resolve the result into something whose structure is consistent with that of its smooth counterpart. Most importantly, the algorithm will generally construct a one-dimensional subcomplex J of K in which edges have positive integer multiplicities. We will prove that J can be unfolded into a union of disjoint closed curves in which all edges have multiplicity one. In other words, we can indeed think of $\mathbb{J} = |J|$ as the limit of a series of smoothly embedded 1-manifolds.

Edge selection. We compute $\mathbb{J}$ by tracing the critical points of the 1-parameter family of functions $h_\lambda = f + \lambda g$. In the piecewise linear limit, the set of critical points is generically a collection of vertices. With varying λ, the critical points move to other vertices and, in the limit, that movement happens along the edges of K and at infinite speed. Instead of keeping track of the critical points and their movements, we construct $\mathbb{J}$ as the union of edges along which the critical points move. In other words, we decide for each edge how many critical points move from one endpoint to the other, and we let J be the collection of edges for which this number is positive. We then make J a subcomplex by adding the endpoints of the edges to the set. Note that we do not have to repeat the construction for the family $\lambda f + g$. By definition, the *Jacobi set* is the underlying space of this subcomplex: $\mathbb{J} = |J|$.

Let uv be an edge of K, and let $\lambda = \lambda_{uv}$ be the moment at which the function values at u and v are the same. At this moment, we have

$h_\lambda(u) = h_\lambda(v)$ and therefore $\lambda_{uv} = [f(v) - f(u)]/[g(u) - g(v)]$. In order for a critical point to travel along the edge from u to v or vice versa, the entire edge must be critical for $\lambda = \lambda_{uv}$. We express this condition by considering the link of u, which is a triangulation of the $(d-1)$-sphere. Let l be the restriction of $h_{\lambda_{uv}}$ to that link.

CRITICAL EDGE LEMMA. *The edge uv belongs to J iff v is a critical point of l. Moreover, the multiplicity of uv in J is the multiplicity of v as a critical point of l.*

We describe shortly how we decide whether or not a vertex is critical and, if it is critical, how we compute its multiplicity. For both operations, it suffices to look at the lower link of v in the link of u. Note that the link of v in the link of u is the same as the link of uv in K, and that l is defined at all vertices of this link as well as at v. We can therefore construct the lower link of uv as the subcomplex induced by the vertices w whose function values are smaller than that of v. We denote this lower link as $\underline{\text{Lk}}\, uv$ and note that it is the same as the lower link of v in the link of u. This formulation makes it obvious that the test of the edge uv is symmetric in u and v. We implement the Critical Edge Lemma as follows:

    integer ISJACOBI(**Edge** uv)
        $\lambda = [f(v) - f(u)]/[g(u) - g(v)]$;
        $\underline{\text{Lk}}\, uv = \{\tau \in \text{Lk}\, uv \mid w \leq \tau \implies h_\lambda(w) < h_\lambda(v)\}$;
        return ISCRITICAL($\underline{\text{Lk}}\, uv$).

As described shortly, Function ISCRITICAL returns 0 if the lower link of uv is that of a regular point, and it returns the multiplicity of the criticality otherwise.

Criticality and multiplicity. The smooth analogue of a vertex link is a sufficiently small sphere drawn around a point x of the manifold $\mathbb{M}$. If $\mathbb{M}$ is a d-manifold the dimension of that sphere is $d-1$. The analogue of the lower link is the portion of that sphere where the value of the function is less than or equal to the value at x. assuming smoothness, the topology of this portion is the same for all sufficiently small spheres, and assuming a Morse function, it has either the homotopy type of a point or that of a sphere. In the former case, x is regular, and in the latter case, x is critical. The dimension of the sphere is one less than the index of x. This suggests we use the reduced Betti numbers of the lower link to classify vertices in a triangulation as shown in Table 2.1. The

	$\tilde{\beta}_{-1}$	$\tilde{\beta}_0$	$\tilde{\beta}_1$	$\tilde{\beta}_2$	...
regular	0	0	0	0	...
0	1	0	0	0	...
1	0	1	0	0	...
2	0	0	1	0	...
3	0	0	0	1	...
...	...	...	...	...	...

Table 2.1. *Classification of vertices into regular points and simple critical points (identified by their index) using the reduced Betti numbers of the lower link.*

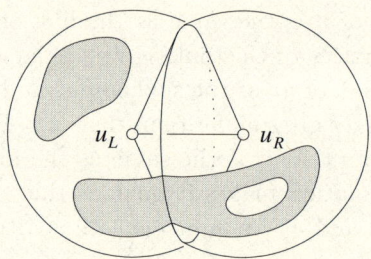

Fig. 2.4. A 2-sphere with two (shaded) oceans and two continents is cut to form two 2-spheres, whose structures are those of a simple index-1 critical point on the left and a simple index-2 critical point on the right.

multiplicity of a vertex u is the sum of reduced Betti numbers of its lower link: $\mu(u) = \sum_{k \geq 0} \tilde{\beta}_{k-1}$. Table 2.1 shows only the regular point, which has multiplicity 0, and the simple critical points, which have multiplicity 1. All other points may be thought of as accumulations of $\mu \geq 2$ simple critical points. In small dimensions, it is easy to effectively unfold such a point into μ simple critical points. We illustrate this for a vertex u in the triangulation of a 3-manifold. Its link is a 2-sphere. We refer to the components of the lower link as *oceans* and to the components of the complement as *continents*. Because the multiplicity is at least 2, we have $\mu(u) = \tilde{\beta}_0 + \tilde{\beta}_1 \geq 2$. There are $\tilde{\beta}_0 + 1$ oceans and $\tilde{\beta}_1 + 1$ continents. As illustrated in Fig. 2.4, we cut the link along a circle that passes through exactly one ocean and one continent and meets each in a single interval.

Next we replace u by two vertices u_L and u_R and connect them to the respective side of the link and to each other. We have $\tilde{\beta}_{0L} + \tilde{\beta}_{0R} = \tilde{\beta}_0$ and $\tilde{\beta}_{1L} + \tilde{\beta}_{1R} = \tilde{\beta}_1$. To guarantee progress, we cut such that $\tilde{\beta}_{0L} + \tilde{\beta}_{1L}$ and $\tilde{\beta}_{0R} + \tilde{\beta}_{1R}$ are both less than μ. This implies that both are at least one, which is equivalent to avoiding the creation of regular points and of minima and maxima. We repeat the process until all vertices are simple critical points. This happens after $\tilde{\beta}_0 + \tilde{\beta}_1 - 1$ splits, which generate $\tilde{\beta}_0$ simple critical points of index 1 and $\tilde{\beta}_1$ simple critical points of index 2. In summary, we compute the multiplicity of a vertex by summing the reduced Betti numbers of its lower link:

```
integer ISCRITICAL(Complex L)
    foreach k ≥ 0 do compute β̃_{k-1} of L endfor;
    return ∑_{k≥0} β̃_{k-1}.
```

To compute the reduced Betti numbers, we may use the Smith normal form algorithm described in [17]. For coefficients modulo 2, this amounts to doing Gaussian elimination on the incidence matrices, which takes a time cubic in the number of simplices in L. Even for integer coefficients, the worst-case running time is polynomial in the number of simplices [13]. For manifolds of dimension $d \leq 5$, we get the lower links as subcomplexes of spheres of dimension $d - 2 \leq 3$. For these we can use the significantly faster incremental algorithm of [7], whose running time is ever so slightly larger than proportional to the number of simplices in the link.

Post-processing. The second step amounts to unfolding the union of edges to a 1-manifold. Define the *degree* of a vertex u as the number of edges in J that share u, where we count each edge with its multiplicity. We use the fact that every vertex has even degree. This is suggested by our analysis of $\mathbb{J}$ in the smooth case but needs a direct proof, which is given below using an elementary parity argument. For good piecewise linear approximations of smooth functions, most vertices will have degree zero. If u has degree 2 or higher, we glue the incident selected edges in pairs. For $d = 2$, this is done so that glued pairs do not cross at u.

EVEN DEGREE LEMMA. *The degree of every vertex in J is even.*

Proof We consider a vertex u and the family of functions $h_\lambda = f + \lambda g$. For $\lambda = \pm\infty$, $\mu(u)$ is independent of f and the same at both extremes. We increase λ continuously from $-\infty$ to $+\infty$. Each time we pass a value $\lambda = \lambda_{uv}$, the status of the neighbor v changes from inside to outside the lower link of u, or vice versa. The effect of this change on the type of u depends on the type of v in the restriction of h_λ to Lk u. Specifically, the multiplicity of u either increases or decreases by the multiplicity of v in the lower link of u. For example, if v is regular then the change has no effect, and if v is a simple critical point then it either increases or decreases $\mu(u)$ by one. The number of operations, each counted with multiplicity of v, is equal to the degree of u. Since we start and end with the same multiplicity of u, we add and subtract equally often, which implies that the degree is even. □

2.5 Jacobi sets of $k+1$ functions

We generalize the results of §2.3 and §2.4 from two to three or more functions defined on the same manifold.

Smooth case. Let $\mathbb{M}$ be a smooth d-manifold and choose a Riemannian metric. Consider $k + 1 \leq d$ Morse functions $f_0, f_1, \ldots, f_k$ on $\mathbb{M}$, and write $\Phi = (f_0, f_1, \ldots, f_k) : \mathbb{M} \to \mathbb{R}^{k+1}$. The generic intersection of the level sets of $j \leq d$ of the f_i is a $(d-j)$-dimensional smooth manifold. Let $t = (t_0, \ldots, \hat{t}_\ell, \ldots, t_k)$ be a k-vector of image values, where the hat indicates that t_ℓ is not in the vector. Generically, the intersection of the corresponding level sets is a smooth $(d-k)$-manifold: $\mathbb{M}_t = \bigcap_{i \neq \ell} f_i^{-1}(t_i)$, and the restriction $f_{\ell t}$ of f_ℓ to $\mathbb{M}_t$ is a Morse function. We call a critical point of this restriction a *simultaneous critical point* of Φ. The *Jacobi set* $\mathbb{J} = \mathbb{J}(\Phi)$ is the closure of the set of simultaneous critical points:

$$\mathbb{J} = \mathrm{cl}\,\{x \in \mathbb{M} \mid x \text{ is critical point of } f_{\ell t}\}, \qquad (2.4)$$

for some index ℓ and some k-vector t. Generically, the k gradients ∇f_i at a point $x \in \mathbb{M}_t$ span the k-dimensional linear subspace of vectors normal to $\mathbb{M}_t$ at x, and x is a critical point of $f_{\ell t}$ iff $\nabla f_\ell(x)$ belongs to this linear subspace. This means that a direct definition of the Jacobi set is:

$$\mathbb{J} = \{x \in \mathbb{M} \mid \mathrm{rank}\,\mathrm{d}\Phi_x \leq k\}, \qquad (2.5)$$

where $\mathrm{d}\Phi_x$ is the d-by-$(k+1)$ matrix whose columns are the gradients of f_0 to f_k at x. Equation (2.5) is symmetric in the components of Φ, which implies that $\mathbb{J}$ is independent of the ordering of the f_i. The linear dependence of the $k+1$ gradient vectors implies that for each $x \in \mathbb{J}$ there is a gradient that can be written as a linear combination of the others: $\nabla f_\ell(x) + \sum_{i \neq \ell} \lambda_i \nabla f_i(x) = 0$. Since the combination of gradients is the gradient of the corresponding combination of functions, this implies

$$\mathbb{J} = \{x \in \mathbb{M} \mid x \text{ is crit. pt. of } f_\ell + \sum_{i \neq \ell} \lambda_i f_i\}, \qquad (2.6)$$

for some index ℓ and some parameters λ_i. Generically, $\mathbb{J}$ is swept out by a k-parameter family of discrete points. It follows that $\mathbb{J}$ is a set of dimension k. Unfortunately, $\mathbb{J}$ is not always a submanifold of $\mathbb{M}$. Following [11], we define $S_r = \{x \in \mathbb{M} \mid \operatorname{rank} d\Phi_x = k + 1 - r\}$. Clearly, $\mathbb{J}$ is the disjoint union of the S_r for $r \geq 1$. The Transversality Theorem explained in §2.2 and the Thom Transversality Theorem [11] page 54 imply that S_r is a submanifold of $\mathbb{M}$ of co-dimension $r(r + d - k - 1)$ for each r, not necessarily a closed submanifold, however. Furthermore the closure of each S_r is the union $S_r \cup S_{r+1} \cup \ldots$, however it is not necessarily a manifold. In particular, $\mathbb{J} = \operatorname{cl} S_1$ is a manifold whenever S_2 (and therefore every S_r for $r \geq 2$) is empty. This happens as long as $d < r(r + d - k - 1)$, for $r = 2$, since a negative co-dimension implies it is empty. This inequality is equivalent to $d > 2k - 2$, and the first time it fails is for a map from a 4-manifold to $\mathbb{R}^4$.

Algorithm. Let K be a triangulation of the d-manifold $\mathbb{M}$ and let $\varphi_i :$ Vert $K \to \mathbb{R}$ be a function defined at the vertices, for $0 \leq i \leq k$. We obtain $f_i : |K| \to \mathbb{R}$ as the piecewise linear extension of φ_i, for each i. As in the case of $k + 1 = 2$ functions, we think of the f_i as limits of smooth functions and construct a k-dimensional subcomplex $J \subseteq K$ as the limit of the Jacobi sets of these smooth functions. Specifically, we compute the multiplicity of every k-simplex in K and define J as the collection of k-simplices with positive multiplicity plus the collection of faces of these k-simplices. By definition, the *Jacobi set* is the underlying space of this subcomplex: $\mathbb{J} = |J|$.

As suggested by (2.6), we introduce a k-parameter family of functions $h_\lambda = f_\ell + \sum_{i \neq \ell} \lambda_i f_i$, where $\lambda = [\lambda_0, \ldots, \hat{\lambda}_\ell, \ldots, \lambda_k]$. Next, we consider a k-simplex σ with vertices $u_0, u_1, \ldots, u_k$ in K and let λ_σ be the k-vector such that $h_{\lambda_\sigma}(u_0) = h_{\lambda_\sigma}(u_1) = \ldots = h_{\lambda_\sigma}(u_k)$. The link of σ is a triangulated $(d - k - 1)$-sphere, and as before we construct the lower link as the subcomplex induced by the vertices w with function values smaller than those of the vertices of σ. Whether or not σ belongs to J depends on the topology or, more precisely, the reduced Betti numbers of the lower link:

```
integer ISJACOBI(k-Simplex σ)
    compute λ = λ_σ;
    Lk σ = {τ ∈ Lk σ | w ≤ τ ⟹ h_λ(w) < h_λ(u_0)};
    return ISCRITICAL(Lk σ).
```

Function ISCRITICAL is the same as in §2.4. It is possibly surprising that the criticality test gets easier the more functions we have, simply because the dimension of the link decreases with increasing k.

2.6 Towards applications

Different applications provide data in different ways and require different adaptations of the algorithm. To illustrate the broad potential of the results presented in this paper, we discuss a few applications while emphasizing the diversity of questions they raise.

Contours. Let $\mathbb{M}$ be a smoothly embedded 2-manifold in $\mathbb{R}^3$ and $a \in \mathbb{S}^2$ an arbitrary but fixed viewing direction. The *contour* is the set of points $x \in \mathbb{M}$ for which the viewing direction belongs to the tangent plane: $a \in \mathrm{T}\mathbb{M}_x$. We introduce two functions, $f, g : \mathbb{M} \to \mathbb{R}$, defined by $f(x) = \langle x, b \rangle$ and $g(x) = \langle x, c \rangle$, where $b, c \in \mathbb{S}^2$ are directions orthogonal to a. Assuming $c \neq \pm b$, $f(x)$ and $g(x)$ are coordinates of the projection of x along the viewing direction. As illustrated in Fig. 2.1, the Jacobi set of f and g is exactly the contour of $\mathbb{M}$, and the projection of the contour is the fold of the two maps. Note that $\mathbb{J} = \mathbb{J}(f, g)$ is a smoothly embedded 1-manifold in $\mathbb{M}$, whereas the fold can have singularities such as cusps and self-intersections.

Suppose $\mathbb{M}$ is given by an approximating simplicial complex K in $\mathbb{R}^3$. The algorithm described in §2.4 amounts to selecting the edges in K for which the plane parallel to a that passes through the edge has both incident triangles on one side. It has been observed experimentally that $\mathbb{J}$ approximates the contour geometrically but not necessarily topologically. Specifically, we can improve the geometric accuracy of the approximation by subdividing K. In the process, spurious cycles of the Jacobi set get smaller but they may not disappear, not even in the limit [5]. This 'topological noise' is an artifact of the piecewise linear approximation and particularly common in hyperbolic regions, such as around the inner longitudinal circle of the torus in Fig. 2.1. We pose the extension of the methods in [8] for measuring topological noise to Jacobi sets as an open problem.

Protein interaction. An energy potential is a type of smooth function common in the sciences. The corresponding force relates to the potential like the gradient relates to the smooth function. As an example, consider the forces that are studied in the context of protein interaction, such as electrostatics and van der Waals. The van der Waals potential decays rapidly, possibly proportional to the sixth power of the distance to the source, while the electrostatic potential decays much slower, proportional to the distance. It is thus believed that the electrostatic potential influences the interaction at an early stage, while the proteins are relatively far apart, by steering them towards each other. Once in close contact, the van der Waals force takes over, which may be the reason why stable interactions require a good amount of local shape complementarity [10]. It would be interesting to put this hypothesis to a computational test in which we can visualize the relationship between different energy potentials through their Jacobi sets. Maybe the hypothesis is more true for some and less for other proteins. On top of visualizing the relationship, it would be useful to quantify the agreement and disagreement between the potentials. It is not entirely clear how to go about constructing such a test. The natural domain for the mentioned potentials is $\mathbb{R}^3$, but it might be more convenient to use two-dimensional domains, such as molecular surfaces [6] and interfaces between proteins [2].

Solar system. Similar to the electrostatic potential, the gravitational potential exerted by a heavenly body decays at a rate that is proportional to the distance. At any moment in time, the gravitational force acting on a point in our solar system depends on its mass and the distance from the sun, the planets and the moons. Let $\mathbb{M} = \mathbb{R}^3 \times \mathbb{R}$ represent space-time in our solar system and let $g : \mathbb{M} \to \mathbb{R}$ be the gravitational potential obtained by adding the contributions of the sun, the planets and the moons. The function is smooth except at the centers of the bodies, where g goes to infinity. At a fixed time t, we have a map from $\mathbb{R}^3$ to $\mathbb{R}$, which generically has four types of critical points. These points trace out curves, which we model as the Jacobi set of g and a second function f that maps every space-time point $(x, t) \in \mathbb{R}^3 \times \mathbb{R}$ to its time: $f(x,t) = t$. Every level set of f is the gravitational potential at some fixed moment in time, so it should be obvious that $\mathbb{J}(f,g)$ is indeed the 1-manifold traced out by the critical points.

The Lagrange points used by NASA in planning the flight paths of their space-crafts are related to these curves but more complicated because they are defined in terms of the interaction between the gravitational force and the momentum of the moving space-craft. It would be interesting to see whether the Lagrange points can also be modeled using the framework of Jacobi sets.

2.7 Discussion

The main contribution of this paper is an algorithm that constructs the Jacobi set of a collection of piecewise linear continuous functions on a common triangulated manifold. The crucial concept in the definition of the Jacobi set is the notion of a critical point of a piecewise linear function. Our decision to use reduced Betti numbers was in part guided by computational convenience and feasibility. The weaker definition based on the Euler characteristic of the lower link used by Banchoff [3] is also possible but misses important portions of the set. The stronger definition that requires the thickened lower link of a regular point be homeomorphic to a ball leads to an undecidable recognition problem for links of dimension 5 or higher [14].

We conclude this section with a list of open questions raised by the work presented in this paper. This list does not include the questions stated in earlier sections.

- The algorithm in §2.4 may be applied more generally than just to manifolds, such as for example to homology manifolds. By definition, these cannot be distinguished from manifolds if we classify links using homology. What is the most general class of simplicial complexes for which our algorithm is meaningful?
- Initial computational experiments indicate that coarse triangulations lead to poor approximations of the Jacobi set. This suggests we need an adaptive meshing method that locally refines the triangulation depending on the available information on the Jacobi set.
- Definition (2.5) of Jacobi sets extends to the case of $k + 1 > d$ Morse functions. According to [11], $\mathbb{J}$ has co-dimension $k - d + 2$ and thus dimension $2d - k - 2$. What is the significance of the Jacobi set of $k + 1 > d$ Morse functions for scientific applications?
- Finally, it would be interesting to explore whether the methods of this paper can be extended to more general vector fields, in particular to smooth vector fields that are not gradient fields.

References

[1] Alexandrov, P.S. (1998). *Combinatorial Topology* (Dover, Mineola, New York).
[2] Ban, Y.-E. A., Edelsbrunner, H. and Rudolph, J. (2002). A definition of interfaces for protein oligomers (Manuscript, Dept. Comput. Sci., Duke Univ., Durham, North Carolina).
[3] Banchoff, T.F. (1970). Critical points and curvature for embedded polyhedral surfaces, *Am. Math. Monthly* **77**, 475–485.
[4] Barnett, T.P., Pierce, D.W. and Schnur, R. (2001). Detection of climate change in the world's oceans, *Science* **292**, 270–274.

[5] Biermann, H., Kristjansson, D. and Zorin, D. (2001). Approximate boolean operations on free-form solids, *Computer Graphics, Proc. SIGGRAPH*, 185–194.
[6] Connolly, M.L. (1983). Analytic molecular surface calculation, *J. Appl. Crystallogr* **6**, 548–558.
[7] Delfinado, C.J.A. and Edelsbrunner, H. (1995). An incremental algorithm for Betti numbers of simplicial complexes on the 3-sphere, *Comput. Aided Geom. Design* **12**, 771–784.
[8] Edelsbrunner, H., Letscher, D. and Zomorodian, A. (2002). Topological persistence and simplification, *Discrete Comput. Geom.* **28**, 511–533.
[9] Edelsbrunner, H and Mücke, E.P. (1990). Simulation of simplicity: a technique to cope with degenerate cases in geometric algorithms, *ACM Trans. Graphics* **9**, 66–104.
[10] Elcock, A.H., Sept, D. and McCammon, J.A. (2001). Computer simulation of protein-protein interactions, *J. Phys. Chem.* **105**, 1504–1518.
[11] Golubitsky, M. and Guillemin, V. (1973). *Stable Mappings and Their Singularities* (Springer-Verlag, New York).
[12] Guillemin, V. and Pollack, A. (1974). *Differential Topology* (Prentice-Hall, Englewood Cliffs, New Jersey).
[13] Kannan, R. and Bachem, A. (1979). Polynomial algorithms for computing the Smith and Hermite normal forms of an integer matrix. *SIAM J. Comput.* **8**, 499–507.
[14] Markov, A.A. (1958). Insolubility of the problem of homeomorphy (in Russian), *Proc. Int. Congress Math.* (Cambridge Univ. Press), 14–21.
[15] Matsumoto, Y. (2002). *An Introduction to Morse Theory*, Amer. Math. Soc. (Translated from Japanese by Hudson, K. and Saito, M.).
[16] Milnor, J. (1963). *Morse Theory* (Princeton Univ. Press, New Jersey).
[17] Munkres, J.R. (1984). *Elements of Algebraic Topology* (Addison-Wesley, Redwood City, California).
[18] Szebehely, V.G. and Mark, H. (1998). *Adventures in Celestial Mechanics*, Second edition (Wiley, New York).
[19] Whitney, H. (1955). On singularities of mappings of Euclidean spaces I: Mappings of the plane to the plane, *Ann. Math.* **62**, 374–410.
[20] Wolpert, N. (2002). *An Exact and Efficient Approach for Computing a Cell in an Arrangement of Quadrics*, Ph. D. thesis (Univ. d. Saarlandes, Saarbrücken, Germany).

3

Approximation of boundary element operators by adaptive $\mathcal{H}^2$-matrices

Steffen Börm
Max-Planck-Institut für Mathematik in den Naturwissenschaften
Inselstraße 22–26, 04103 Leipzig
Email: sbo@mis.mpg.de

Wolfgang Hackbusch
Max-Planck-Institut für Mathematik in den Naturwissenschaften
Inselstraße 22–26, 04103 Leipzig
Email: wh@mis.mpg.de

Abstract

The discretization of integral operators corresponding to non-local kernel functions typically gives rise to densely populated matrices. In order to be able to treat these matrices in an efficient manner, they have to be compressed, e.g., by panel clustering algorithms, multipole expansions or wavelet techniques.

By choosing the correct panel clustering approach, the resulting approximation of the matrix can be written in the form of a so-called $\mathcal{H}^2$-matrix. The $\mathcal{H}^2$-matrix representation can be computed for fairly general kernel functions by a black box algorithm that requires only pointwise evaluations of the kernel function.

Although this technique leads to good results, the expansion system tends to contain a certain level of redundancy that leads to an unnecessarily high complexity for the memory requirements and the matrix-vector multiplication. We present two variants of the original method that can compress the matrix even further. Both methods work on the fly, i.e., it is not necessary to keep the original $\mathcal{H}^2$-matrix in memory, and both methods perform an algebraic compression, so that the black box character of the algorithm is preserved.

3. Approximation of boundary element operators

3.1 Introduction

3.1.1 Model problem

We consider integral operators of the form

$$K[u](x) = \int_\Gamma \kappa(x,y)u(y)\,dy \qquad (x \in \Gamma), \tag{3.1}$$

where u is a suitable function defined on the boundary Γ of a domain $\Omega \subseteq \mathbb{R}^d$, where $\kappa(\cdot,\cdot)$ is a kernel function defined on $\mathbb{R}^d \times \mathbb{R}^d$, possibly with a singularity at the diagonal $\{(x,x) : x \in \mathbb{R}^d\}$. Discretizing this operator by a Galerkin method leads to a matrix $\mathbf{K} \in \mathbb{R}^{\mathfrak{I} \times \mathfrak{I}}$ defined by

$$\mathbf{K}_{ij} := \int_\Gamma \int_\Gamma \kappa(x,y) \Phi_j(y) \Phi_i(x)\,dy\,dx \qquad (i,j \in \mathfrak{I}), \tag{3.2}$$

where $(\Phi_i)_{i \in \mathfrak{I}}$ is the set of basis functions. Typical kernel function $\kappa(\cdot,\cdot)$ (e.g., from BEM applications) have non-local support, therefore the matrix $\mathbf{K}$ will be densely populated. If we store $\mathbf{K}$ in the typical two-dimensional array, we will need N^2 units of memory ($N := \#\mathfrak{I}$), and obviously this complexity is not acceptable for high-dimensional problems.

3.1.2 Compression techniques

There are different techniques for reducing the complexity: We can use the fast Fourier transform to diagonalize $\mathbf{K}$ if this matrix has Toeplitz structure. This leads to a complexity of $\mathcal{O}(N \log N)$, but the Toeplitz structure occurs only in special situations.

Due to these restrictions, more general techniques have been developed that replace the matrix $\mathbf{K}$ by data-sparse approximations $\tilde{\mathbf{K}}$ having a complexity of $\mathcal{O}(N \log^\lambda N)$ (where λ depends on the choice of the method): The panel clustering method [9] replaces the kernel function locally by separable functions, e.g., Taylor expansions or polynomial interpolants. The multipole approach [10] uses a more sophisticated expansion of the kernel function that is more efficient, but requires the analytical expansion formulae for each kernel under consideration. Finally, we can use the wavelet compression technique [5] provided we are willing to use wavelet discretizations and can construct suitable wavelet spaces for the domain Γ.

The main advantage of the panel clustering technique is that it is relatively simple, that it can be applied to a large number of practical problems and that it can be implemented as a black box method. The main disadvantage is that the standard implementations are based on an approximation of the kernel function in d-dimensional subdomains, even if Γ is only $(d-1)$-dimensional.

If we want to perform not only matrix-vector multiplications, but also matrix-matrix additions, multiplications or even the inversion of matrices efficiently, most of the techniques mentioned before can not be applied directly. The method of hierarchical matrices ($\mathcal{H}$-matrices) [6, 7, 1] generalizes the concept of separable expansions in order to find a representation of matrices that makes it possible to perform the sophisticated arithmetic operations mentioned above with complexity $\mathcal{O}(N \log^\lambda N)$, where λ depends on the type of operation. $\mathcal{H}^2$-matrices [8, 3, 4] are a refinement of $\mathcal{H}$-matrices that introduce an additional hierarchical structure in order to reach the *optimal complexity* $\mathcal{O}(N)$ for the matrix-vector multiplication.

3.1.3 Adaptive panel clustering method

Our goal is to find a panel clustering algorithm that "automatically" finds improved expansion systems without sacrificing the black box character and general applicability of the original method.

The idea is to start with a representation of the boundary element matrix by an $\mathcal{H}^2$-matrix constructed by means of polynomial interpolation of the kernel function and then apply an algebraic optimization that removes superfluous functions from the expansion system while controlling the approximation error. Due to this additional optimization, setting up the optimized $\mathcal{H}^2$-matrix has a higher complexity than in the standard case, but the complexity of the matrix-vector multiplication is reduced significantly.

Another advantage of using standard interpolation as the basis of our method is that the total approximation error of the adaptive method is a combination of the well-known interpolation error and the user-defined error bound of the algebraic optimization procedure.

3.1.4 Organization of this paper

This paper is organized in five sections: In the current section, we will introduce a model problem and discuss some of the related algorithms. The next section is devoted to the definition of the $\mathcal{H}^2$-approximation of the matrix corresponding to the model problem. The third section describes two techniques for compressing the original $\mathcal{H}^2$-approximation, and the fourth section contains numerical experiments that demonstrate that the compression rates achieved by our techniques are much better than those of standard methods.

3.2 Approximation of integral operators by $\mathcal{H}^2$-matrices

In this section, we give a short introduction to $\mathcal{H}^2$-matrices and a simple method for using them to approximate matrices corresponding to integral operators (cf. [3]).

3.2.1 Interpolation

Due to reasons that will become clear later on, we cannot hope to find a global approximation of the matrix $\mathbf{K}$. Therefore we consider only a submatrix corresponding to a block $\tau \times \sigma$, where $\tau, \sigma \subseteq \mathcal{J}$ (recall that $\mathcal{J}$ is the index set corresponding to the finite element space). Let $B^\tau, B^\sigma \subseteq \mathbb{R}^d$ be axis-parallel boxes satisfying

$$\operatorname{supp} \Phi_i \subseteq B^\tau, \quad \operatorname{supp} \Phi_j \subseteq B^\sigma \qquad (i \in \tau, j \in \sigma).$$

The boxes B^τ and B^σ will be called the *bounding boxes* corresponding to τ and σ.

Now we need to find a separable approximation of the kernel function $\kappa(\cdot, \cdot)$ on $B^\tau \times B^\sigma$. The simplest possible approach is to use interpolation: We fix an m-th order d-dimensional interpolation operator

$$\mathcal{I} : C([-1,1]^d) \to \mathcal{Q}^{d,m},$$

where $\mathcal{Q}^{d,m}$ denotes the set of d-dimensional polynomials of order m. For any d-dimensional axis-parallel box B, we introduce the transformed interpolation operator

$$\mathcal{I}^B : C(B) \to \mathcal{Q}^{d,m}, \quad u \mapsto (\mathcal{I}[u \circ \Psi]) \circ \Psi^{-1},$$

where Ψ is the standard affine mapping from $[-1,1]^d$ to B.

The approximation of $\kappa(\cdot, \cdot)$ on the domain $B^\tau \times B^\sigma$ is given by

$$\tilde{\kappa}^{\tau,\sigma} := (\mathcal{I}^{B^\tau} \otimes \mathcal{I}^{B^\sigma})\kappa \in \mathcal{Q}^{2d,m}.$$

If we denote the interpolation points corresponding to $\mathcal{I}^{B^\tau}$ and $\mathcal{I}^{B^\sigma}$ by $(x_\nu^\tau)_{\nu \in K}$ and $(x_\mu^\sigma)_{\mu \in K}$ and the corresponding Lagrange polynomials by

$(\mathcal{L}_\nu^\tau)_{\nu \in K}$ and $(\mathcal{L}_\mu^\sigma)_{\mu \in K}$, we have

$$\mathcal{I}^{B^\tau} u = \sum_{\nu \in K} u(x_\nu^\tau) \mathcal{L}_\nu^\tau, \quad \mathcal{I}^{B^\sigma} v = \sum_{\mu \in K} v(x_\mu^\sigma) \mathcal{L}_\mu^\sigma$$

and therefore

$$\tilde{\kappa}^{\tau,\sigma}(x,y) = ((\mathcal{I}^{B^\tau} \otimes \mathcal{I}^{B^\sigma})\kappa)(x,y) = \sum_{\nu \in K} \sum_{\mu \in K} \kappa(x_\nu^\tau, x_\mu^\sigma) \mathcal{L}_\nu^\tau(x) \mathcal{L}_\mu^\sigma(y),$$

i.e., in $\tilde{\kappa}^{\tau,\sigma}(\cdot,\cdot)$, we have found a separable approximation of $\kappa(\cdot,\cdot)$. Replacing $\kappa(x,y)$ by $\tilde{\kappa}^{\tau,\sigma}(x,y)$ in (3.2) leads to the approximated matrix entries

$$\tilde{\mathbf{K}}_{ij} := \int_\Gamma \int_\Gamma \tilde{\kappa}^{1,2}(x,y) \Phi_j(y) \Phi_i(x) \, dy \, dx$$

$$= \sum_{\nu \in K} \sum_{\mu \in K} \kappa(x_\nu^1, x_\mu^2) \int_\Gamma \mathcal{L}_\nu^1(x) \Phi_i(x) \, dx \int_\Gamma \mathcal{L}_\mu^2(y) \Phi_j(y) \, dy$$

$$= (\mathbf{V}^1 \mathbf{S}^{1,2} \mathbf{V}^{2\top})_{ij} \tag{3.3}$$

for $i \in \tau$ and $j \in \sigma$, where

$$\mathbf{V}_{i\nu}^1 := \int_\Gamma \mathcal{L}_\nu^1(x) \Phi_i(x), \quad \mathbf{V}_{j\mu}^2 := \int_\Gamma \mathcal{L}_\mu^2(y) \Phi_j(y) \text{ and } \mathbf{S}_{\nu\mu}^{1,2} := \kappa(x_\nu^1, x_\mu^2).$$

This factorized form can be evaluated in $\mathcal{O}(k\#\tau + k\#\sigma + k^2)$ operations for $k := \#K$ and is therefore much more efficient than the standard form if k is significantly smaller than N.

3.2.2 Approximation error

Let us now take a look at the error introduced by replacing κ by its approximation $\tilde{\kappa}^{1,2}$. For tensor product interpolation, error estimates of the form

$$\|\kappa - \tilde{\kappa}^{1,2}\|_{\infty, B^1 \times B^2} \leq C_{\text{in}}(m) \frac{c_1^m}{(m+1)!} \operatorname{diam}(B^1 \times B^2)^{m+1}$$

$$\times \sum_{p=1}^{2d} \|\partial_p^{m+1} \kappa\|_{\infty, B^1 \times B^2} \tag{3.4}$$

hold, where $c_1 \in \mathbb{R}_{>0}$ is a constant and $C_{\text{in}}(m)$ is a polynomial in m (cf. [3]). In order to make use of this estimate, we need a bound for the derivatives of κ. Typical kernel functions are *asymptotically smooth*, i.e., they have a singularity of order $g \in \mathbb{N}_0$ at $x = y$ and satisfy the inequality

$$|\partial_x^\alpha \partial_y^\beta \kappa(x,y)| \leq C_{\text{apx}}(\alpha+\beta)!(c_0\|x-y\|)^{-g-|\alpha|-|\beta|} \qquad (3.5)$$

for some constants $C_{\text{apx}}, c_0 \in \mathbb{R}_{>0}$. Combining this inequality with (3.4), we find

$$\|\kappa - \tilde{\kappa}^{\tau,\sigma}\|_{\infty, B^\tau \times B^\sigma}$$
$$\leq C_{\text{in}}(m) C_{\text{apx}} \left(\frac{c_1 \operatorname{diam}(B^\tau \times B^\sigma)}{c_0 \operatorname{dist}(B^\tau, B^\sigma)} \right)^{m+1} \operatorname{dist}(B^\tau, B^\sigma)^{-g}.$$

This estimate implies that we can expect good convergence only if the diameter of $B^\tau \times B^\sigma$ can be bounded by the distance of the boxes, i.e., if B^τ and B^σ satisfy an *admissibility condition* of the type

$$\operatorname{diam}(B^\tau \times B^\sigma) \leq \eta \operatorname{dist}(B^\tau, B^\sigma) \qquad (3.6)$$

for some parameter $\eta \in \mathbb{R}_{>0}$. If this inequality holds, we find

$$\|\kappa - \tilde{\kappa}^{\tau,\sigma}\|_{\infty, B^\tau \times B^\sigma} \leq C_{\text{in}}(m) C_{\text{apx}} \left(\frac{c_1 \eta}{c_0} \right)^{m+1} \operatorname{dist}(B^\tau, B^\sigma)^{-g}, \qquad (3.7)$$

i.e., we have exponential convergence in m if $\eta < c_0/c_1$ holds.

Remark 3.1 *The Newton kernel $\kappa(x,y) = 1/\|x-y\|$ satisfies (3.5) with $c_0 = 1$. Tensor product Chebyshev interpolation satisfies (3.4) with $c_1 = 1/4$. This implies that $\eta \in (0, 4)$ guarantees exponential convergence of the kernel approximation.*

3.2.3 Local approximation

The admissibility condition (3.6) implies $\operatorname{dist}(B^\tau, B^\sigma) > 0$, so we cannot expect to find a global approximation of the form (3.3) for the entire matrix. Instead, we split the matrix into suitable blocks and approximate each block separately.

In order to get an efficient algorithm, we use a hierarchical approach to construct the block decomposition: We organize the degrees of freedom in the form of a *cluster tree*, i.e., a tree with root $\mathfrak{J}$ and the property that if a node $\tau \subseteq \mathfrak{J}$ is not a leaf, it is the disjoint union of all of its sons. The nodes of a cluster tree will be called *clusters*.

For a given cluster tree $\mathcal{T}_\mathfrak{J}$, we denote the set of clusters by $\mathcal{T}_\mathfrak{J}$ and the set of sons for a given cluster $\tau \in \mathcal{T}_\mathfrak{J}$ by $\operatorname{sons}(\tau)$. We fix a bounding box B^τ for each cluster $\tau \in \mathcal{T}_\mathfrak{J}$.

We can use the admissibility condition (3.6) in combination with the cluster tree to construct the desired partition of the matrix: A pair (τ, σ) is called *admissible*, if condition (3.6) holds. We start with $(\mathfrak{I}, \mathfrak{I})$ and split a cluster pair as long as it is not admissible. This leads to the following algorithm:

```
procedure subdivide(τ, σ, var P);
begin
    if (τ,σ) is admissible then P := P ∪ {(τ,σ)}
    else if sons(τ) = ∅ or sons(σ) = ∅ then P := P ∪ {(τ,σ)}
    else
        for τ' ∈ sons(τ) do
            for σ' ∈ sons(σ) do
                subdivide(τ', σ', P)
end;

procedure divide(var P);
begin
    P := ∅; subdivide(ℑ, ℑ, P)
end;
```

This algorithm gives us a set P satisfying

$$\mathfrak{I} \times \mathfrak{I} = \bigcup_{(\tau,\sigma) \in P} \tau \times \sigma,$$

i.e., a partition of the index set $\mathfrak{I} \times \mathfrak{I}$ corresponding to the matrix $\mathbf{K}$. Obviously, an entry (τ, σ) can only appear in P if either it is admissible or if τ or σ is a leaf. This distinction is represented by the splitting

$$P_{\text{far}} := \{(\tau, \sigma) \in P : (\tau, \sigma) \text{ is admissible}\}, \quad P_{\text{near}} := P \setminus P_{\text{far}}$$

of P into admissible and non-admissible blocks. The non-admissible blocks are stored without compression, while we apply our approximation scheme to the admissible blocks.

3.2.4 Compressed representation

For each cluster τ, we denote the interpolation points and Lagrange polynomials corresponding to $\mathcal{I}^{B^\tau}$ by $(x_\nu^\tau)_{\nu \in K}$ and $(\mathcal{L}_\nu^\tau)_{\nu \in K}$ and introduce the matrix $\mathbf{V}^\tau \in \mathbb{R}^{\tau \times K}$ by setting

$$\mathbf{V}_{i\nu}^\tau := \int_\Gamma \mathcal{L}_\nu^\tau(x) \Phi_i(x) \, dx \qquad (i \in \tau, \nu \in K). \tag{3.8}$$

The family $(\mathbf{V}^\tau)_{\tau \in T_\mathfrak{J}}$ is called a *cluster basis*. Let $(\tau, \sigma) \in P_{\text{far}}$. We replace the kernel function κ by its interpolant

$$\tilde{\kappa}^{\tau,\sigma} := (\mathcal{I}^{B^\tau} \otimes \mathcal{I}^{B^\sigma})\kappa = \sum_{\nu \in K} \sum_{\mu \in K} \kappa(x_\nu^\tau, x_\mu^\sigma) \mathcal{L}_\nu^\tau(x) \mathcal{L}_\mu^\sigma(y)$$

and get the approximate matrix $\tilde{\mathbf{K}}^{\tau,\sigma} \in \mathbb{R}^{\tau \times \sigma}$ defined by

$$\begin{aligned}\tilde{\mathbf{K}}_{ij}^{\tau,\sigma} &:= \sum_{\nu \in K} \sum_{\mu \in K} \kappa(x_\nu^\tau, x_\mu^\sigma) \int_\Gamma \mathcal{L}_\nu^\tau(x) \Phi_i(x)\, dx \int_\Gamma \mathcal{L}_\mu^\sigma(y) \Phi_j(y)\, dy \\ &= (\mathbf{V}^\tau \mathbf{S}^{\tau,\sigma} \mathbf{V}^{\sigma\top})_{ij} \end{aligned} \qquad (3.9)$$

for $i \in \tau$, $j \in \sigma$, where $\mathbf{S}^{\tau,\sigma} \in \mathbb{R}^{K \times K}$ is given by

$$\mathbf{S}_{\nu,\mu}^{\tau,\sigma} := \kappa(x_\nu^\tau, x_\mu^\sigma) \qquad (\nu, \mu \in K). \qquad (3.10)$$

The approximation $\tilde{\mathbf{K}} \in \mathbb{R}^{\mathfrak{J} \times \mathfrak{J}}$ of the matrix $\mathbf{K}$ is given by

$$\tilde{\mathbf{K}}_{ij} := \begin{cases} (\mathbf{V}^\tau \mathbf{S}^{\tau,\sigma} \mathbf{V}^{\sigma\top})_{ij} & \text{if } (i,j) \in \tau \times \sigma \text{ for } (\tau, \sigma) \in P_{\text{far}}, \\ \mathbf{K}_{ij} & \text{otherwise,} \end{cases} \quad (i, j \in \mathfrak{J}).$$

3.2.5 Fast matrix-vector multiplication

Now that we have derived a compact approximation of the matrix $\mathbf{K}$, we consider the efficient computation of the product of $\tilde{\mathbf{K}}$ and a given vector $u \in \mathbb{R}^\mathfrak{J}$.

The straightforward method is to loop over all blocks $(\tau, \sigma) \in P$ and multiply them by u. This approach is not optimal, since we have to perform the multiplication by $\mathbf{V}^{\sigma\top}$ for each single block of the form $(\tau, \sigma) \in P_{\text{far}}$. In order to remove this redundancy, we split the matrix-vector multiplication into four parts:

(i) **Forward transformation:** Compute $\hat{u}_\sigma := \mathbf{V}^{\sigma\top} u|_\sigma$ for all $\sigma \in T_\mathfrak{J}$.

(ii) **Multiplication:** Compute $\hat{v}_\tau := \sum_{\sigma, (\tau,\sigma) \in P_{\text{far}}} \mathbf{S}^{\tau,\sigma} \hat{u}_\sigma$ for all $\tau \in T_\mathfrak{J}$.

(iii) **Backward transformation:** Initialize the output vector v by zero and add up the contributions of all clusters: $v|_\tau := v|_\tau + \mathbf{V}^\tau \hat{v}_\tau$.

(iv) **Nearfield:** Add the uncompressed parts: $v|_\tau := \mathbf{K}|_{\tau \times \sigma} u|_\sigma$ for all $(\tau, \sigma) \in P_{\text{near}}$.

This approach is more efficient than the naive method, but can be improved even further: Let $\tau \in T_\mathfrak{J}$ be a node that is not a leaf. Let $\tau' \in \text{sons}(\tau)$. Since our interpolation operators are projections onto

$Q^{d,m}$, we find that
$$\mathcal{L}_\nu^\tau = \mathcal{I}^{B^{\tau'}} \mathcal{L}_\nu^\tau = \sum_{\nu' \in K} \mathcal{L}_\nu^\tau(x_{\nu'}^{\tau'}) \mathcal{L}_{\nu'}^{\tau'} \qquad \text{holds for all } \nu \in K.$$

For a given index $i \in \tau'$, this implies
$$\mathbf{V}_{i\nu}^\tau = \int_\Gamma \mathcal{L}_\nu^\tau(x) \Phi_i(x)\, dx$$
$$= \sum_{\nu' \in K} \mathcal{L}_\nu^\tau(x_{\nu'}^{\tau'}) \int_\Gamma \mathcal{L}_{\nu'}^{\tau'}(x) \Phi_i(x)\, dx = (\mathbf{V}^{\tau'} \mathbf{B}^{\tau',\tau})_{i\nu}, \quad (3.11)$$

where $\mathbf{B}^{\tau',\tau} \in \mathbb{R}^{K \times K}$ is defined by
$$\mathbf{B}_{\nu'\nu}^{\tau',\tau} := \mathcal{L}_\nu^\tau(x_{\nu'}^{\tau'}) \qquad (\nu, \nu' \in K). \tag{3.12}$$

The equation (3.11) describes an essential property of the basis: The restriction of $\mathbf{V}^\tau$ to the subset τ' belongs to the range of $\mathbf{V}^{\tau'}$, the cluster bases are *nested*.

Using this property, we can derive recursive procedures for performing the first and third step of the matrix-vector multiplication:

```
procedure fastforward(σ, u, var (û_σ)_{σ∈T_J});
begin
    if sons(σ) = ∅ then û_σ := V^σ ᵀ u|_σ
    else
        for σ' ∈ sons(σ) do begin
            fastforward(σ', u, û);
            û_σ := û_σ + B^{σ',σ} ᵀ û_{σ'}
        end
end;

procedure fastbackward(τ, var v, (v̂_τ)_{τ∈T_J});
begin
    if sons(τ) = ∅ then v|_τ := V^τ v̂_τ
    else
        for τ' ∈ sons(τ) do begin
            v̂_{τ'} := v̂_{τ'} + B^{τ',τ} v̂_τ;
            fastbackward(τ', v, v̂)
        end
end;
```

In order to stress the similarities of both procedures, we have not included the necessary initialization: Before calling the fast forward transformation, the output coefficients $(\hat{u}_\sigma)_{\sigma \in T_\mathcal{J}}$ have to be set to zero.

The advantage of the recursive procedures is that we have to store the matrices $\mathbf{V}^\tau$ only for the leaves of the cluster tree. For all other clusters, it is sufficient to store the small transfer matrices $\mathbf{B}^{\tau',\tau}$. This leads to a significant reduction in the storage complexity.

Remark 3.2 (Complexity) *In typical situations, building the matrices $\mathbf{B}^{\tau',\tau}$, $\mathbf{V}^\tau$ and $\mathbf{S}^{\tau,\sigma}$ can be done in $\mathcal{O}(Nm^d)$ operations. The matrix-vector multiplication based on the recursive procedures requires $\mathcal{O}(Nm^d)$ operations, too (cf. [3]).*

3.3 Orthonormalization

Our approximation is created by a d-dimensional interpolation operator, but enters the computations only in the form of boundary integrals over a $(d-1)$-dimensional surface. Therefore we can expect that the used expansion system is too rich and that it should be possible to construct reduced expansion systems, e.g., harmonic polynomials if $\kappa(\cdot,\cdot)$ is the kernel function corresponding to the Laplace equation. We do this by applying algebraic algorithms to the original $\mathcal{H}^2$-matrix.

The algebraic equivalent of the continuous expansion system is the cluster basis $(\mathbf{V}^\tau)_{\tau \in T_\mathcal{J}}$ (cf. (3.8)), which can be considered to represent the Galerkin discretizations of the expansion functions.

Therefore, our goal is to find a reduced cluster basis $(\tilde{\mathbf{V}}^\tau)_{\tau \in T_\mathcal{J}}$ and then to find an approximation of the original $\mathcal{H}^2$-matrix in terms of the new basis.

3.3.1 Orthonormalized cluster basis

Let us fix a cluster $\tau \in T_\mathcal{J}$, and let $\nu \in K$. If the basis function $\mathcal{L}_\nu^\tau$ is redundant, its Galerkin projection can be represented in terms of the projections of the remaining basis functions, therefore we can represent the ν-th column of $\mathbf{V}^\tau$ in terms of the other columns, and this implies that $\mathbf{V}^\tau$ is rank-deficient.

This suggests a simple method for the elimination of redundant expansion functions: We try to orthonormalize $\mathbf{V}^\tau$, i.e., to find a rank $\tilde{k}^\tau \in \mathbb{N}$ and matrix $\mathbf{Q}^\tau \in \mathbb{R}^{K \times \tilde{k}^\tau}$ such that $\tilde{\mathbf{V}}^\tau := \mathbf{V}^\tau \mathbf{Q}^\tau \in \mathbb{R}^{\tau \times \tilde{k}^\tau}$ is orthogonal.

The orthogonality of $\tilde{\mathbf{V}}^\tau$ is equivalent to

$$\mathbf{I} = \tilde{\mathbf{V}}^{\tau\top} \tilde{\mathbf{V}}^\tau = \mathbf{Q}^{\tau\top}(\mathbf{V}^{\tau\top}\mathbf{V}^\tau)\mathbf{Q}^\tau. \tag{3.13}$$

A matrix $\mathbf{Q}^\tau$ satisfying this condition can be found by different algorithms. We use the Schur decomposition

$$\mathbf{P}^\top \mathbf{G} \mathbf{P} = \mathbf{D}$$

of the positive semidefinite symmetric matrix $\mathbf{G} := \mathbf{V}^\tau{}^\top \mathbf{V}^\tau$, where $\mathbf{P} \in \mathbb{R}^{K \times k}$ is a square orthogonal matrix (recall that K is the index set corresponding to the original expansion system and k is its cardinality), $\mathbf{D} \in \mathbb{R}^{k \times k}$ is diagonal and the entries of $\mathbf{D}$ are ordered in a monotonously increasing sequence.

If all the diagonal entries of D were positive, we could set $\mathbf{Q}^\tau := \mathbf{P}\mathbf{D}^{-1/2}$ and find

$$\mathbf{Q}^\tau{}^\top(\mathbf{V}^\tau{}^\top\mathbf{V}^\tau)\mathbf{Q}^\tau = \mathbf{D}^{-1/2}\mathbf{P}^\top\mathbf{G}\mathbf{P}\mathbf{D}^{-1/2} = \mathbf{D}^{-1/2}\mathbf{D}\mathbf{D}^{-1/2} = \mathbf{I},$$

so we would have found a solution for (3.13).

For rank-deficient matrices $\mathbf{V}^\tau$, $\mathbf{D}$ will have zero entries, so we have to modify our approach. The idea is to choose those entries of $\mathbf{D}$ that are larger than a given threshold $\epsilon \in \mathbb{R}_{>0}$: We set

$$\tilde{k}^\tau := \min\{l \in \{1, \ldots, k\} \ : \ \sum_{p=l+1}^{k} \mathbf{D}_{pp} \leq \epsilon\}$$

and use $\mathbf{D}^\dagger \in \mathbb{R}^{k \times \tilde{k}^\tau}$ given by

$$\mathbf{D}^\dagger_{ij} = \begin{cases} \mathbf{D}_{ij}^{-1/2} & \text{if } i = j, \\ 0 & \text{otherwise} \end{cases} \quad (i \in \{1, \ldots, k\}, j \in \{1, \ldots, \tilde{k}^\tau\})$$

in order to define

$$\tilde{\mathbf{V}}^\tau := \mathbf{V}^\tau \mathbf{P} \mathbf{D}^\dagger.$$

This implies

$$\tilde{\mathbf{V}}^\tau \tilde{\mathbf{V}}^\tau{}^\top = \mathbf{V}^\tau \mathbf{P}(\mathbf{D}^\dagger)^2 \mathbf{P}^\top \mathbf{V}^\tau{}^\top$$

and therefore

$$\mathbf{V}^\tau{}^\top \tilde{\mathbf{V}}^\tau \tilde{\mathbf{V}}^\tau{}^\top \mathbf{V}^\tau = \mathbf{G}\mathbf{P}(\mathbf{D}^\dagger)^2\mathbf{P}^\top\mathbf{G} = \mathbf{P}\mathbf{D}(\mathbf{D}^\dagger)^2\mathbf{D}\mathbf{P}^\top.$$

Due to orthogonality, the best approximation of $\mathbf{V}^\tau$ in the range of $\tilde{\mathbf{V}}^\tau$ is given by $\tilde{\mathbf{V}}^\tau \tilde{\mathbf{V}}^\tau{}^\top \mathbf{V}^\tau$ and satisfies the error estimate

$$\begin{aligned}\|\mathbf{V}^\tau - \tilde{\mathbf{V}}^\tau \tilde{\mathbf{V}}^\tau{}^\top \mathbf{V}^\tau\|_F^2 &= \operatorname{tr}(\mathbf{V}^\tau{}^\top\mathbf{V}^\tau - \mathbf{V}^\tau{}^\top\tilde{\mathbf{V}}^\tau\tilde{\mathbf{V}}^\tau{}^\top\mathbf{V}^\tau) \\ &= \operatorname{tr}(\mathbf{P}(\mathbf{D} - \mathbf{D}(\mathbf{D}^\dagger)^2\mathbf{D})\mathbf{P}^\top) \\ &= \operatorname{tr}(\mathbf{D} - \mathbf{D}(\mathbf{D}^\dagger)^2\mathbf{D}) \leq \epsilon.\end{aligned}$$

3.3.2 Nested basis

Applying the straightforward orthonormalization described above to all clusters $\tau \in T_\mathfrak{J}$ will give us a reduced basis, but this basis will no longer be nested, i.e., (3.11) will no longer hold, so we would not be able to use the fast matrix-vector multiplication algorithm.

Therefore we have to introduce a further modification: Let $\tau \in T_\mathfrak{J}$ be a cluster with $\text{sons}(\tau) \neq \emptyset$. Due to (3.11), we have

$$\mathbf{V}^\tau|_{\tau' \times K} = \mathbf{V}^{\tau'} \mathbf{B}^{\tau',\tau} \qquad \text{for all } \tau' \in \text{sons}(\tau). \tag{3.14}$$

We want to find a reduced matrix $\tilde{\mathbf{V}}^\tau$ satisfying a similar equation for reduced transfer matrices $\tilde{\mathbf{B}}^{\tau',\tau}$.

Suppose we have already computed $\tilde{\mathbf{V}}^{\tau'}$ for all $\tau' \in \text{sons}(\tau)$ of τ. We approximate $\mathbf{V}^{\tau'}$ in (3.14) in terms of $\tilde{\mathbf{V}}^{\tau'}$, i.e., we apply the orthogonal projection to the range of $\tilde{\mathbf{V}}^{\tau'}$ to both sides of the equation:

$$\tilde{\mathbf{V}}^{\tau'} \tilde{\mathbf{V}}^{\tau'^\top} \mathbf{V}^\tau|_{\tau' \times K} = \tilde{\mathbf{V}}^{\tau'} \tilde{\mathbf{V}}^{\tau'^\top} \mathbf{V}^{\tau'} \mathbf{B}^{\tau',\tau}.$$

Let $\mathbf{W}^{\tau'} := \tilde{\mathbf{V}}^{\tau'^\top} \mathbf{V}^{\tau'}$, and let $\widehat{\mathbf{V}}^\tau \in \mathbb{R}^{\tau \times K}$ be defined by

$$\widehat{\mathbf{V}}^\tau|_{\tau' \times K} := \tilde{\mathbf{V}}^{\tau'} \mathbf{W}^{\tau'} \mathbf{B}^{\tau',\tau} \tag{3.15}$$

for all $\tau' \in \text{sons}(\tau)$. $\widehat{\mathbf{V}}^\tau$ is well-defined since the sons $\tau' \in \text{sons}(\tau)$ are disjoint and their union is τ.

The matrix $\widehat{\mathbf{V}}^\tau$ is the orthogonal projection of $\mathbf{V}^\tau$ onto the space spanned by the ranges of the matrices $\tilde{\mathbf{V}}^{\tau'}$ corresponding to the sons of τ, i.e. $\widehat{\mathbf{V}}^\tau$ is the best approximation of $\mathbf{V}^\tau$ we can get without giving up nestedness.

Now we can apply the same orthogonalization procedure as before to $\widehat{\mathbf{V}}^\tau$ instead of $\mathbf{V}^\tau$ in order to find a matrix $\mathbf{Q}^\tau$ satisfying

$$\mathbf{I} = \tilde{\mathbf{V}}^{\tau^\top} \tilde{\mathbf{V}}^\tau = \mathbf{Q}^{\tau^\top} (\widehat{\mathbf{V}}^{\tau^\top} \widehat{\mathbf{V}}^\tau) \mathbf{Q}^\tau.$$

The matrix $\tilde{\mathbf{V}}^\tau$ is given by

$$\tilde{\mathbf{V}}^\tau := \widehat{\mathbf{V}}^\tau \mathbf{Q}^\tau,$$

and this implies

$$\tilde{\mathbf{V}}^\tau|_{\tau' \times \tilde{k}^\tau} = \widehat{\mathbf{V}}^\tau|_{\tau' \times K} \mathbf{Q}^\tau = \tilde{\mathbf{V}}^{\tau'} \mathbf{W}^{\tau'} \mathbf{B}^{\tau',\tau} \mathbf{Q}^\tau = \tilde{\mathbf{V}}^{\tau'} \tilde{\mathbf{B}}^{\tau',\tau}$$

for $\tilde{\mathbf{B}}^{\tau',\tau} := \mathbf{W}^{\tau'} \mathbf{B}^{\tau',\tau} \mathbf{Q}^\tau$, i.e., the new cluster basis $(\tilde{\mathbf{V}}^\tau)_{\tau \in T_\mathfrak{J}}$ is nested (cf. (3.11)).

The Gram matrix $\widehat{\mathbf{V}}^{\tau^\top}\widehat{\mathbf{V}}^\tau$ used in the computation of $\tilde{\mathbf{B}}^{\tau',\tau}$ can be constructed by means of the equation

$$\widehat{\mathbf{V}}^{\tau^\top}\widehat{\mathbf{V}}^\tau = \sum_{\tau' \in \text{sons}(\tau)} \mathbf{B}^{\tau',\tau^\top} \mathbf{W}^{\tau'^\top} \tilde{\mathbf{V}}^{\tau'^\top} \tilde{\mathbf{V}}^{\tau'} \mathbf{W}^{\tau'} \mathbf{B}^{\tau',\tau}$$

$$= \sum_{\tau' \in \text{sons}(\tau)} \mathbf{B}^{\tau',\tau^\top} \mathbf{W}^{\tau'^\top} \mathbf{W}^{\tau'} \mathbf{B}^{\tau',\tau},$$

due to the orthogonality of $\tilde{\mathbf{V}}^{\tau'}$.

By splitting the matrix $\mathbf{W}^\tau$ and using the nestedness of the bases $(\mathbf{V}^\tau)_{\tau \in T_\mathcal{I}}$ and $(\tilde{\mathbf{V}}^\tau)_{\tau \in T_\mathcal{I}}$, we find that the matrix $\mathbf{W}^\tau$ can be represented in the form

$$\mathbf{W}^\tau = \tilde{\mathbf{V}}^{\tau^\top} \mathbf{V}^\tau = \sum_{\tau' \in \text{sons}(\tau)} \tilde{\mathbf{B}}^{\tau',\tau^\top} \tilde{\mathbf{V}}^{\tau'^\top} \mathbf{V}^{\tau'} \mathbf{B}^{\tau',\tau}$$

$$= \sum_{\tau' \in \text{sons}(\tau)} \tilde{\mathbf{B}}^{\tau',\tau^\top} \mathbf{W}^{\tau'} \mathbf{B}^{\tau',\tau}; \tag{3.16}$$

Hence this computation requires only the matrices $\mathbf{W}^{\tau'}$ corresponding to the sons τ' of τ and the transfer matrices $\mathbf{B}^{\tau',\tau}$ and $\tilde{\mathbf{B}}^{\tau',\tau}$.

The following algorithm computes the matrices $\tilde{\mathbf{V}}^\tau$ for leaves $\tau \in T_\mathcal{I}$ and the transfer matrices $\tilde{\mathbf{B}}^{\tau',\tau}$ for the remaining clusters $\tau \in T_\mathcal{I}$ with $\tau' \in \text{sons}(\tau)$:

procedure orthonormalize(τ);
begin
   if sons(τ) = $\emptyset$ then begin
     $\mathbf{G} := \mathbf{V}^{\tau^\top}\mathbf{V}^\tau$;   {*Build Gram matrix*}
     Find $\mathbf{Q}^\tau$ with $\mathbf{Q}^{\tau^\top}\mathbf{G}\mathbf{Q}^\tau = \mathbf{I}$;
     $\tilde{\mathbf{V}}^\tau := \mathbf{V}^\tau \mathbf{Q}^\tau$;   {*New basis for τ*}
     $\mathbf{W}^\tau := \tilde{\mathbf{V}}^{\tau^\top}\mathbf{V}^\tau$ {*Update transformation matrix*}
   end else begin
     for $\tau' \in \text{sons}(\tau)$ do orthonormalize(τ');   {*Recursion*}
     $\mathbf{G} := 0$;   {*Build projected Gram matrix*}
     for $\tau' \in \text{sons}(\tau)$ do $\mathbf{G} := \mathbf{G} + \mathbf{B}^{\tau',\tau^\top} \mathbf{W}^{\tau'^\top} \mathbf{W}^{\tau'} \mathbf{B}^{\tau',\tau}$;
     Find $\mathbf{Q}^\tau$ with $\mathbf{Q}^{\tau^\top}\mathbf{G}\mathbf{Q}^\tau = \mathbf{I}$;
     for $\tau' \in \text{sons}(\tau)$ do $\tilde{\mathbf{B}}^{\tau',\tau} := \mathbf{W}^{\tau'}\mathbf{B}^{\tau',\tau}\mathbf{Q}^\tau$;   {*New basis for τ*}
     $\mathbf{W}^\tau := 0$;   {*Update transformation matrix*}
     for $\tau' \in \text{sons}(\tau)$ do $\mathbf{W}^\tau := \mathbf{W}^\tau + \tilde{\mathbf{B}}^{\tau',\tau^\top} \mathbf{W}^{\tau'} \mathbf{B}^{\tau',\tau}$
   end
end;

This procedure is local, i.e., only the matrix $\mathbf{V}^\tau$ is needed for the computation if τ is a leaf, and only the matrices $\mathbf{B}^{\tau',\tau}$ for $\tau' \in \text{sons}(\tau)$ are needed if τ is not a leaf. By using temporary variables that are initialized at the beginning of the procedure (cf. (3.8) and (3.12)), we do not need to store the entire original cluster basis.

3.3.3 Conversion of the coefficient matrices

Let $(\tau, \sigma) \in P_{\text{far}}$. The submatrix corresponding to this block of an $\mathcal{H}^2$-matrix is given in the form $\mathbf{V}^\tau \mathbf{S}^{\tau,\sigma} {\mathbf{V}^\sigma}^\top$ (cf. (3.9)).

In order to find a representation of this block with respect to the new cluster basis $(\tilde{\mathbf{V}}^\tau)_{\tau \in T_\mathcal{J}}$, we once more use the orthogonal projection and get

$$\tilde{\mathbf{V}}^\tau {\tilde{\mathbf{V}}^\tau}^\top \mathbf{V}^\tau \mathbf{S}^{\tau,\sigma} {\mathbf{V}^\sigma}^\top \tilde{\mathbf{V}}^\sigma {\tilde{\mathbf{V}}^\sigma}^\top = \tilde{\mathbf{V}}^\tau \tilde{\mathbf{S}}^{\tau,\sigma} {\tilde{\mathbf{V}}^\sigma}^\top$$

as the best approximation with

$$\tilde{\mathbf{S}}^{\tau,\sigma} := {\tilde{\mathbf{V}}^\tau}^\top \mathbf{V}^\tau \mathbf{S}^{\tau,\sigma} {\mathbf{V}^\sigma}^\top \tilde{\mathbf{V}}^\sigma = \mathbf{W}^\tau \mathbf{S}^{\tau,\sigma} {\mathbf{W}^\sigma}^\top,$$

so we need only the matrices $(\mathbf{W}^\tau)_{\tau \in T_\mathcal{J}}$ computed as a byproduct in the basis orthonormalization algorithm.

3.3.4 Complete algebraic recompression

The orthogonalization algorithm considers only the expansion system itself, but not the kernel function we intend to approximate. Therefore we can expect improved results if we include the coefficient matrices $\mathbf{S}^{\tau,\sigma}$ in addition to the cluster basis.

This is done by the algorithm introduced in [2] for dense and hierarchical matrices. Since $\mathcal{H}^2$-matrices are a specialization of hierarchical matrices, we could convert the $\mathcal{H}^2$-matrix representation into the form of an $\mathcal{H}$-matrix representation and apply the algorithm from [2] directly. For the conversion, we would have to compute the matrices $(\mathbf{V}^\tau)_{\tau \in T_\mathcal{J}}$ for *all* clusters $\tau \in T_\mathcal{J}$, not only for the leaves, i.e., we could not benefit from the more compact recursive representation based on (3.11).

Instead of computing the matrices $(\mathbf{V}^\tau)_{\tau \in T_\mathcal{J}}$ in advance and using them to expand the factorized form $\mathbf{V}^\tau \mathbf{S}^{\tau,\sigma} {\mathbf{V}^\sigma}^\top$ of the $\mathcal{H}^2$-matrix blocks to the form used in the $\mathcal{H}$-matrix conversion algorithm, we keep the blocks factorized as long as possible and expand them during the course of the recursion: We use the set of blocks related to ancestors of τ given by

$$A^\tau := \{\sigma \in T_\mathcal{J} \; : \; \exists \tau_0 \in T_\mathcal{J} : \tau \subseteq \tau_0, (\tau_0, \sigma) \in P_{\text{far}}\}$$

and store the intermediate results of the transformation of these blocks in a family $(\mathbf{C}^{\tau,\sigma})_{\sigma \in A^\tau}$ of auxiliary matrices. After a suitable basis for a cluster has been found, all the blocks are transformed into the new basis and the results are stored in another family $(\widehat{\mathbf{C}}^{\tau,\sigma})_{\sigma \in A^\tau}$ of auxiliary matrices. This leads to the following algorithm:

```
procedure recompression(τ, Aᵀ);
begin
    if sons(τ) = ∅ then for σ ∈ Aᵀ do Ĉᵀ,σ := Cᵀ,σ      {Leaves: No conversion}
    else begin
        Let sons(τ) = {τ₁,...,τₛ};
        for i := 1 to s do begin                         {Compute ancestor blocks}
            Rᵀⁱ := {σ ∈ T₃ : (τᵢ,σ) ∈ P_far}; Aᵀⁱ := Aᵀ ∪ Rᵀⁱ;
            for σ ∈ Aᵀ do Cᵀⁱ,σ := Bᵀⁱ,ᵀ Cᵀ,σ;
            for σ ∈ Rᵀⁱ do Cᵀⁱ,σ := Sᵀⁱ,σ;
            recompression(τᵢ, Aᵀⁱ)                       {Determine cluster basis for sons}
        end;
        for i := 1 to s do                               {Compute Gram matrix for all sons}
            for j := 1 to s do begin
                Gᵢⱼ := 0;
                for σ ∈ Aᵀ do Gᵢⱼ := Gᵢⱼ + Ĉᵀⁱ,σ ᵀ Tσ Ĉᵀʲ,σ
            end;
```

Find an orthogonal $\tilde{k}^\tau$-column matrix $\mathbf{Q} = (\mathbf{Q}_1^\top, \ldots, \mathbf{Q}_s^\top)^\top$ maximizing

$$\left\| (\mathbf{Q}_1^\top, \ldots, \mathbf{Q}_s^\top) \begin{pmatrix} \mathbf{G}_{11} & \cdots & \mathbf{G}_{1s} \\ \vdots & \ddots & \vdots \\ \mathbf{G}_{s1} & \cdots & \mathbf{G}_{ss} \end{pmatrix} \right\|_F.$$

```
        for i := 1 do s do begin                         {Convert blocks to the new basis}
            B̃ᵀⁱ,ᵀ := Qᵢ;
            for σ ∈ Aᵀ do Ĉᵀ,σ := B̃ᵀ ᵀ Cᵀ,σ
        end
    end
end;
```

The matrices $(\mathbf{T}^\sigma)_{\sigma \in T_3}$ appearing in this procedure are defined by $\mathbf{T}^\sigma := \mathbf{V}^{\sigma\top} \mathbf{V}^\sigma$ and can be computed by a recursion similar to (3.16).

3.4 Numerical Experiments

For our experiments, we consider the kernel function

$$\kappa(x, y) := \frac{1}{4\pi \|x - y\|}$$

corresponding to the three-dimensional single layer potential operator. The domain Ω is the unit ball in $\mathbb{R}^3$ and therefore Γ is the unit sphere in three dimensions.

3. Approximation of boundary element operators

N	original		orthogonalized		recompressed	
	Build/s	MVM/s	Build/s	MVM/s	Build/s	MVM/s
512	4.33	0.07	5.42	0.01	6.17	0.01
2048	20.16	0.48	25.83	0.09	30.46	0.04
8192	83.50	1.71	108.24	0.45	223.51	0.24
32768	333.25	6.87	435.54	1.78	1163.62	1.02
131072	1315.24	27.05	1718.66	7.42	5683.39	4.08
524288					27366.03	16.07

Table 3.1. *Time in seconds for setup and matrix-vector multiplication*

We approximate Γ by a regular triangulation consisting of plane triangles and use piecewise constant basis functions for our Galerkin discretization.

The original discretization is performed for an interpolation order of 4, and we choose $\epsilon = 10^{-4}$ as the threshold for the algebraic compression algorithm.

We will first consider the speed of our algorithms. Table 3.1 lists the time needed[1] for building the different $\mathcal{H}^2$-matrix approximations and for performing the matrix-vector multiplications:

We can see that both compression techniques lead to a significant reduction in the time for the matrix-vector multiplication: For the simple orthogonalization algorithm the matrix-vector multiplication is speeded up by a factor of almost 4, while the full recompression even gives us a factor of more than 6.

On the other hand, building the fully recompressed $\mathcal{H}^2$-matrix requires much more time than building the $\mathcal{H}^2$-matrix with orthogonalized cluster basis. This is not surprising, since the full recompression algorithm has to consider *all* admissible blocks of the $\mathcal{H}^2$-matrix, while the orthogonalization algorithm works only based on the cluster basis.

In Table 3.2, we consider the amount of memory required for storing the original and compressed $\mathcal{H}^2$-approximations.

The columns "orig", "orth." and "recomp." give the number of bytes of storage needed per degree of freedom. In Table 3.2, the advantage of the full recompression method is obvious: It reduces the memory requirements by more than 95% compared to the original method, while the orthogonalization algorithm reaches only 86%.

---

[1] All computations were performed on Sun Ultra 3cu processors running at 900 MHz.

N	original	orthogonalized	recompressed
512	92108	6633	3527
2048	139020	15057	5015
8192	152422	19873	5848
32768	152618	21235	6171
131072	146278	20345	6138

Table 3.2. *Memory requirement in bytes per degree of freedom*

N	original	orthogonalized	recompressed
512	4.53033_{-5}	4.52851_{-5}	4.53183_{-5}
2048	5.08654_{-5}	5.11205_{-5}	5.10336_{-5}
8192	6.63957_{-5}	6.65225_{-5}	6.64592_{-5}
32768	7.22294_{-5}	7.22293_{-5}	7.23569_{-5}

Table 3.3. *Relative approximation error*

Of course, we are not only interested in fast algorithms, we also need the results to be sufficiently precise. We approximate the operator norm of matrices by starting with a random vector and performing 100 steps of the power iteration. The values of $\|\mathbf{K} - \tilde{\mathbf{K}}\|_2 / \|\mathbf{K}\|_2$ are collected in Table 3.3.

Here, the columns "orig.", "orth." and "recomp." give the norm of the difference between the densely populated matrix and the original $\mathcal{H}^2$-matrix, the $\mathcal{H}^2$-matrix with orthogonalized cluster bases and the recompressed $\mathcal{H}^2$-matrix.

Obviously, the additional errors introduced by orthogonalization and recompression are negligible. This means that both algorithms yield a significant reduction in the computational complexity without affecting the precision. The compression ratio of the full recompression procedure is much better than that of the simple orthogonalization method, but this advantage is complemented by the significantly higher complexity.

References

[1] Börm, S., Grasedyck, L. and Hackbusch, W. (2002). Introduction to hierarchical matrices with applications (Tech. Rep. **18**, Max Planck Institute for Mathematics in the Sciences). To appear in: *Engineering Analysis with Boundary Elements*.

[2] Börm, S. and Hackbusch, W. (2002). Data-sparse approximation by adaptive $\mathcal{H}^2$-matrices, *Computing* **69**, 1–35.
[3] Börm, S. and Hackbusch, W. (2002). $\mathcal{H}^2$-matrix approximation of integral operators by interpolation, *Applied Numerical Mathematics* **43**, 129–143.
[4] Börm, S., Löhndorf, L. and Melenk, J.M. (2002). Approximation of integral operators by variable-order interpolation (Tech. Rep. **82**, Max Planck Institute for Mathematics in the Sciences).
[5] Dahmen, W. and Schneider, R. (1999). Wavelets on manifolds I: Construction and domain decomposition, *SIAM Journal of Mathematical Analysis* **31**, 184–230.
[6] Hackbusch, W. (1999). A sparse matrix arithmetic based on $\mathcal{H}$-matrices. Part I: Introduction to $\mathcal{H}$-matrices, *Computing* **62**, 89–108.
[7] Hackbusch, W. and Khoromskij, B. (2000). A sparse matrix arithmetic based on $\mathcal{H}$-matrices. Part II: Application to multi-dimensional problems, *Computing* **64**, 21–47.
[8] Hackbusch, W., Khoromskij, B. and Sauter, S. (2000) On $\mathcal{H}^2$-matrices, *Lectures on Applied Mathematics* (Bungartz, H., Hoppe, R. and Zenger, C., eds.), 9–29.
[9] Hackbusch, W. and Nowak, Z.P. (1989). On the fast matrix multiplication in the boundary element method by panel clustering, *Numerische Mathematik* **54**, 463–491.
[10] Rokhlin, V. (1958). Rapid solution of integral equations of classical potential theory, *Journal of Computational Physics* **60**, 187–207.

4

Quantum Complexity of Numerical Problems

Stefan Heinrich
Department of Computer Science
University of Kaiserslautern
Germany
Email: heinrich@informatik.uni-kl.de
Homepage: http://www.uni-kl.de/AG-Heinrich

Abstract

A challenging question in the overlap of computer science, mathematics, and physics, is the exploration of potential capabilities of quantum computers. Milestones were the factoring algorithm of Shor (1994) and the search algorithm of Grover (1996). So far, major research was concentrated on discrete and algebraic problems. Much less was known about computational problems of analysis, including such a prominent example as high dimensional numerical integration, which is well-studied in the classical settings. We seek to understand how efficiently this and related problems can be solved in the quantum model of computation (i.e., on a quantum computer) and how the outcome compares to the efficiency of deterministic or randomized algorithms on a classical (i.e. non-quantum) computer. In this paper we give a survey of the state of the art in this field, including also a brief introduction to the general ideas of quantum computing.

4.1 Introduction

A quantum computer is a computing device based on quantum mechanical laws of the (sub)atomic world. The idea of such a computer was developed by Feynman [8] in 1982. He emphasized that simulating quantum mechanics on a classical computer is extremely hard, probably infeasible. So why not try to simulate quantum mechanics using quantum devices themselves. (Thoughts in this direction were also expressed by Manin [21] in 1980, see also [22].) In 1985 Deutsch [6] presented a formal model of computation for quantum computing. A sensational breakthrough of quantum computing was Shor's [29] result of 1994: He produced a polynomial (in the number of bits) algorithm for factoring large integers (no polynomial classical – deterministic or randomized – algorithm is known). Another seminal result is due to Grover [9] in 1996: Given a function $f : \{0,\ldots,N-1\} \to \{0,1\}$ with the property that there is a unique i_0 with $f(i_0) = 1$, the task is to find this i_0. The function is given as a black box, which means that function values are only available at request, by query calls. Classical deterministic or randomized algorithms cannot solve this problem with less than $\Omega(N)$ queries to f. Grover's quantum algorithm needs $\mathcal{O}(\sqrt{N})$ quantum queries (we explain this notion later).

These developments triggered an explosion of efforts in quantum computing. The challenges of quantum computing to physicists are to find quantum systems suitable for computation, i.e., to built a quantum computer. In recent years, various realizations are tested in laboratories, but so far only a small number of system components (qubits) is possible.

What are the challenges of quantum computing to mathematicians and computer scientists? To find more problems for which quantum algorithms are better than all known classical algorithms (Shor's result belongs to this category). And even stronger: Find more problems for which quantum algorithms are **provably** better than all possible classical algorithms (like in Grover's result). In this vein all kinds of discrete problems are being investigated. Furthermore, Feynman's original idea was realized: various quantum algorithms for quantum mechanical simulations were suggested which are better than known classical algorithms (however, no proof of superiority over all possible classical algorithms was given for these types of numerical problems).

The study of numerical problems from the stronger point of view of provable superiority was begun by Brassard, Høyer, Mosca, and Tapp [5, 4]. They exhibited a quantum algorithm for computing the mean of $\{0,1\}$-valued sequences, provably superior to all classical algorithms. That this algorithm is even optimal among all quantum algorithms was shown by Nayak and Wu [23]. First ideas of how to apply quantum algorithms for integration were expressed by Abrams and Williams [1]. Novak [26] carried out a first systematic study of integration, including lower bounds and provable superiority. He considered integration of functions from Hölder spaces. In the sequel this study was continued and widely extended. The author considered the mean of p-summable sequences and integration in L_p [12] and in Sobolev spaces [13]. A quantum complexity theory for continuous problems of numerical analysis, that is, the quantum setting of information-based complexity theory, was developed in [12]. Novak and the author [17] completed the analysis of mean computation for p-summable sequences. Path integration was considered by Traub and Woźniakowski [32]. First approaches towards approximation of functions by quantum algorithms were made by Novak, Sloan, and Woźniakowski [27] for high dimensional Hilbert function classes. The first matching bounds for approximation are given by the author in [15], where the case of Sobolev embeddings is considered.

Complexity of numerical problems is studied in the general framework of information-based complexity theory. By now for many important problems of numerical analysis, including high dimensional integration in various function spaces, matching upper and lower complexity bounds are known for both the classical deterministic and randomized setting. It is a challenging task to study these problems in the setting of quantum computation. Once such results are obtained, one can compare them to the deterministic and randomized classical ones to understand the possible speedups by quantum algorithms.

After a short review of crucial notions from the classical deterministic and randomized setting of information-based complexity theory, we explain basic ideas of quantum computation and present the quantum setting. Then we discuss recent results on quantum algorithms, lower bounds, and complexity for various integration problems, like approximating the mean of p-summable sequences and the integral of functions from Hölder and Sobolev spaces. It turns out that in many cases there is a quadratic speed-up of quantum algorithms over classical randomized ones, and, with increasing dimension of the integration domain, a polynomial speed-up of arbitrarily large degree of quantum algorithms over classical deterministic ones. Finally, we also give an example of a function class in which quantum integration gives an exponential speedup over deterministic algorithms and discuss some recent new directions.

For further reading on quantum computation we recommend the surveys by Aharonov [2], Ekert, Hayden, and Inamori [7], Shor [30], and the monographs by Pittenger [28], Gruska [10], and Nielsen and Chuang [24].

For notions and results in information-based complexity theory see the monographs by Traub, Wasilkowski, and Woźniakowski [31] and Novak [25], and the survey by the author [11] of the randomized setting. For a first overview of quantum complexity theory for numerical problems we refer to Heinrich and Novak [16], while the connections to Monte Carlo algorithms are emphasized in a survey by the author [14].

4.2 Numerical problems in the classical settings

Let D be a non-empty set, F a set of real-valued functions on D, G a normed space, and let

$$S : F \to G \qquad (4.1)$$

be a mapping, which we refer to as the solution operator. $S(f) \in G$ represents the exact solution of our numerical problem in consideration at input $f \in F$.

As an example, consider the computation of the mean (or, equivalently – up to the weighting factor – the summation of finite sequences). Let $D = \{0, \ldots, N-1\}$ and $G = \mathbb{R}$. For a function $f : D \to \mathbb{R}$ define

$$S(f) = S_N f = \frac{1}{N} \sum_{i=0}^{N-1} f(i).$$

To specify the set of inputs, consider the **discrete L_p-classes**: For $1 \leq p \leq \infty$, let $L_p^N = \mathbb{R}^N$, equipped with the norm

$$\|f\|_{L_p^N} = \left(\frac{1}{N} \sum_{i=0}^{N-1} |f(i)|^p \right)^{1/p}$$

if $1 \leq p < \infty$, and

$$\|f\|_{L_\infty^N} = \max_{0 \leq i < N} |f(i)|.$$

Now we let

$$F = \mathcal{B}(L_p^N) = \{ f \in L_p^N \, : \, \|f\|_{L_p^N} \leq 1 \}$$

be the unit ball of L_p^N.

Our second example is multivariate integration. Here we let $d \in \mathbb{N}$, $D = [0,1]^d$, and $G = \mathbb{R}$. For a Lebesgue integrable function $f : [0,1]^d \to \mathbb{R}$ put

$$S(f) = I_d f := \int_{[0,1]^d} f(t) dt.$$

As input sets we consider two basic types of classes:

Hölder classes: For $r \in \mathbb{N}_0$ and $0 < s \leq 1$ define

$$F = \mathcal{B}(F_d^{r,s}) = \{ f \in C^r([0,1]^d), \, \|f\|_\infty \leq 1,$$
$$|\partial^\alpha f(x) - \partial^\alpha f(y)| \leq |x-y|^s, |\alpha| = r \}.$$

Here $C^r([0,1]^d)$ stands for the set of r times continuously differentiable functions.

Sobolev classes: For $r, d \in \mathbb{N}$, and $1 \le p \le \infty$, satisfying $r/d > 1/p$ (the Sobolev embedding condition) let

$$F = \mathcal{B}(W_{p,d}^r) = \{f \in L_p([0,1]^d) : \|\partial^\alpha f\|_{L_p} \le 1, \, |\alpha| \le r\},$$

where ∂^α denotes the weak partial derivative.

In the classical deterministic setting we will consider the following general form of an algorithm which uses n function values:

$$A_n(f) = \varphi(f(x_1), \ldots, f(x_n)).$$

Here $x_i \in D$ $(i = 1, \ldots, n)$ are any points, and $\varphi : \mathbb{R}^n \to G$ is an arbitrary mapping (both chosen by the algorithm designer). The error of A_n over F is defined as

$$e(S, A_n, F) = \sup_{f \in F} \|S(f) - A_n(f)\|_G.$$

The crucial quantity of complexity analysis in this setting is the deterministic n-th minimal error

$$e_n^{\text{det}}(S, F) = \inf_{A_n} e(S, A_n, F),$$

which is the minimal possible error reachable among all algorithms which use at most n function values.

In the classical randomized setting the general form of an algorithm which uses n function values is the following: $A_n = (A_n^\omega)_{\omega \in \Omega}$, where

$$A_n^\omega(f) = \varphi^\omega(f(x_1^\omega), \ldots, f(x_n^\omega)),$$

$(\Omega, \Sigma, \mathbf{P})$ is a probability space, $x_i^\omega \in D$ are random points in D, and $\varphi^\omega : \mathbb{R}^n \to G$ $(\omega \in \Omega)$ is a random mapping. Then the error of A_n is defined as

$$e(S, A_n, F) = \sup_{f \in F} (\mathbf{E}\|S(f) - A_n^\omega(f)\|_G^2)^{1/2}.$$

This leads to the randomized n-th minimal error

$$e_n^{\text{ran}}(S, F) = \inf_{A_n} e(S, A_n, F),$$

that is, the minimal possible (mean-square) error among all randomized algorithms using at most n function values. So far a short overview of the classical settings. For further reading we refer to [31, 25, 11]. In §4.4 we introduce the quantum setting.

4.3 Quantum computation

The mathematical framework for the basic unit of quantum computing is the two dimensional complex Hilbert space $H_1 := \mathbb{C}^2$. The unit sphere of H_1 serves as the state space of quantum systems with two classical states. The classical states correspond to the elements of the unit vector basis $e_0, e_1 \in H_1$. Such systems are called **qubits** – quantum bits. Following quantum mechanics notation, we write $|0\rangle$ instead of e_0 and $|1\rangle$ instead of e_1.

A quantum computer is an m-**qubit system**, that is, a system of m interacting qubits (which can be manipulated, as will be explained below). Such a system is represented by the tensor product

$$H_m := \underbrace{H_1 \otimes H_1 \otimes \cdots \otimes H_1}_{m}.$$

This is the 2^m-dimensional complex Hilbert space, with its canonical basis

$$e_{i_0} \otimes e_{i_1} \otimes \cdots \otimes e_{i_{m-1}} \quad (i_0, i_1, \ldots, i_{m-1}) \in \{0,1\}^m.$$

Let us introduce further notational conventions:

$$e_{i_0} \otimes e_{i_1} \otimes \cdots \otimes e_{i_{m-1}} =: |i_0\rangle |i_1\rangle \ldots |i_{m-1}\rangle$$
$$=: |i\rangle$$

where $i := (i_0 i_1 \ldots i_{m-1})_2 := \sum_{k=0}^{m-1} i_k 2^{m-1-k}$.

The vectors $|i\rangle = |i_0\rangle |i_1\rangle \ldots |i_{m-1}\rangle$ represent the classical states. A general state of the quantum system is given by the superposition

$$|\xi\rangle = \sum_{i=0}^{2^m-1} \alpha_i |i\rangle \quad \left(\sum_{i=0}^{2^m-1} |\alpha_i|^2 = 1\right).$$

(Let us make the following notational conventions connected with the (Dirac) notation of quantum mechanics: if $|\ldots\rangle$ contains a nonnegative integer inside, or a typical symbol denoting such a number, like $i, j, k, \ldots$, we mean the canonical basis vector corresponding to this number, if $|\ldots\rangle$ contains any other symbol, like $|\xi\rangle, |\varphi\rangle, |\psi\rangle$, we mean any vector of H_m).

How to use such m-qubit systems for computing? To explain this at a simple example, let us first go back to classical computations, and consider the addition of two binary numbers

$$(i_0 i_1 \ldots i_{m-1})_2 + (j_0 j_1 \ldots j_{m-1})_2 = (k_0 k_1 \ldots k_m)_2.$$

A classical implementation would look as follows:

$$i_0, \ldots, i_{m-1}, j_0, \ldots, j_{m-1}, 0, \ldots, 0$$
$$\downarrow$$
$$i_0, \ldots, i_{m-1}, j_0, \ldots, j_{m-1}, k_0, \ldots, k_m$$

where the computation of the bits of the sum $k_0, \ldots, k_m$ is realized using circuits of classical gates {and, or, not, xor} in the usual way: add the last bit, then the second last plus the carry bit etc.

How to operate m-qubit quantum systems? Which operations are allowed? Schrödinger's equation implies: all evolutions of a quantum system must be represented by unitary transforms of H_m.

Quantum computing assumes that we are able to perform a number of elementary (quantum) gates on the system

Next we present some common quantum gates. The simplest ones are the **one qubit gates**. They manipulate only one component of the tensor product $H_1 \otimes H_1 \otimes \cdots \otimes H_1$. Formally, a one qubit gate is given by a unitary operators on H_1. The action on the whole tensor product is then obtained by taking the tensor product of this operator with the identities on the other components. An important one qubit gate is the **Hadamard gate** defined by

$$|0\rangle \to \frac{|0\rangle + |1\rangle}{\sqrt{2}}$$
$$|1\rangle \to \frac{|0\rangle - |1\rangle}{\sqrt{2}}$$

(the values on the basis vectors define the unitary transform uniquely). Another example is the family of **phase shifts** given for any fixed $\theta \in [0, 2\pi]$ by

$$\alpha_0 |0\rangle + \alpha_1 |1\rangle \to \alpha_0 |0\rangle + e^{i\theta} \alpha_1 |1\rangle.$$

In a similar way, **two qubit gates** are defined. They manipulate two components of $H_1 \otimes H_1 \otimes \cdots \otimes H_1$. We consider only one example, the **quantum xor gate** (also called **controlled-not gate**): Its unitary action from $H_1 \otimes H_1$ to $H_1 \otimes H_1$ is determined by

$$|0\rangle |0\rangle \to |0\rangle |0\rangle$$
$$|0\rangle |1\rangle \to |0\rangle |1\rangle$$
$$|1\rangle |0\rangle \to |1\rangle |1\rangle$$
$$|1\rangle |1\rangle \to |1\rangle |0\rangle$$

4. Quantum Complexity

That is, if the first bit is zero, nothing happens to the second, and if the first is one, the second is negated (controlled not).

A basic result of quantum computing states that these gates are already enough to approximate any unitary operator on $H_1 \otimes H_1 \otimes \cdots \otimes H_1$; see, e.g. [24]:

The Hadamard gate, the phase shift $\theta = \pi/4$ and the xor gate form an approximately universal system of gates – each unitary transform of H_m can be approximated in the operator norm to each precision by a finite composition of these gates (up to a complex factor).

Hence, if one is able to realize these basic gates in physical systems, one can do quantum computing. Physicists are working on implementations of these gates in various quantum systems such as photons, trapped ions, magnetic resonance systems etc.

Of course, the statement above does not yet say anything about the efficiency of such an approximation, that is, how many of these elementary gates are needed. This is the theme of quantum complexity theory.

Let us come back to our examples and emphasize two important aspects:

1. These gates can transform classical states into superpositions. Example: The Hadamard gate applied to the first and then to the second qubit

$$|0\rangle |0\rangle \longrightarrow \frac{1}{2} \left(|0\rangle |0\rangle + |0\rangle |1\rangle + |1\rangle |0\rangle + |1\rangle |1\rangle \right).$$

2. They act also on superpositions. Examples:
The quantum xor:

$$\alpha_0 |0\rangle |0\rangle + \alpha_1 |0\rangle |1\rangle + \alpha_2 |1\rangle |0\rangle + \alpha_3 |1\rangle |1\rangle$$
$$\downarrow$$
$$\alpha_0 |0\rangle |0\rangle + \alpha_1 |0\rangle |1\rangle + \alpha_2 |1\rangle |1\rangle + \alpha_3 |1\rangle |0\rangle$$

Quantum addition of binary numbers:

$$\sum \alpha_{ij} |i_0\rangle \ldots |i_{m-1}\rangle |j_0\rangle \ldots |j_{m-1}\rangle |0\rangle \ldots |0\rangle$$
$$\downarrow$$
$$\sum \alpha_{ij} |i_0\rangle \ldots |i_{m-1}\rangle |j_0\rangle \ldots |j_{m-1}\rangle |k_0\rangle \ldots |k_m\rangle$$

That is, in the quantum world, we add all possible binary m-digit numbers in parallel.

So is a quantum computer an ideal parallel computer, with exponentially many processors? Not exactly, it is not that easy! The point is that we cannot access all components of the superposition. We have to measure the quantum system, which destroys the superposition.

Quantum computing assumes that we are able to access the results of the quantum computation process via measurement (with respect to the canonical basis).

This means the following: Measuring a system in a (superposition) state
$$|\psi\rangle = \sum_{i=0}^{2^m-1} \alpha_i |i\rangle \quad \left(\sum_{i=0}^{2^m-1} |\alpha_i|^2 = 1\right)$$
results in one of the classical states:
$$|i\rangle \quad \text{with probability} \quad |\alpha_i|^2 \quad (i = 0, \ldots, 2^m - 1).$$

So, returning to our example of binary addition, after measurement we would get just
$$|i_0\rangle \ldots |i_{m-1}\rangle |j_0\rangle \ldots |j_{m-1}\rangle |k_0\rangle \ldots |k_m\rangle$$
with probability $|\alpha_{ij}|^2$. This simple example shows two typical features: A quantum computer is indeed a powerful device due to the exponential parallelism of computation. On the other hand, we cannot "look into" the device to read off all parallel results. To exploit a quantum computer properly, one has to go deeper: Some elaborate algorithmic techniques are needed to transform the quantum state in such a way that the desired final result can be obtained via measurement (with high probability). The quantum Fourier transform, phase estimation, and Grover's iteration are the foremost tools to reach this goal (see the references mentioned above for details). This concludes our short introduction into quantum computing.

4.4 The quantum setting for numerical problems

Now we become more specific and describe the formal quantum model of computation (the general way a quantum algorithm should look like), suited to handle numerical problems. An important issue not discussed so far is this: How does the quantum algorithm get information about $f \in F$? (Remember, the quantum algorithm is supposed to approximate $S(f)$ for each $f \in F$.) Let us first look at the binary case, that is, we are dealing with Boolean functions $f : \{0, 1, \ldots, 2^{m_1} - 1\} \to \{0, 1\}$.

4. Quantum Complexity

The classical (i.e., non-quantum) black box (also called query, or subroutine) maps $(i, 0, k)$ to $(i, f(i), k)$, that is, at request i the subroutine writes $f(i)$ into some memory space originally filled by 0. The entry $k \in \{0, 1, \ldots, 2^{m-m_1-1} - 1\}$ just represents the contents of additional "work bits", which are needed for the rest of the computation and which are not touched by the query. This is the way we always describe numerical algorithms – e.g. in the case of integration, it is tacitly assumed that we have at our disposal the values of the function to integrate. The reflection above leads us to set

$$|i\rangle \, |0\rangle \, |k\rangle \to |i\rangle \, |f(i)\rangle \, |k\rangle.$$

in the quantum case. This is, however, not yet complete – to define a unitary operator $Q_f : H_m \to H_m$ uniquely, we have to extend the mapping above to a bijection of classical states. A standard (and convenient, from the point view of implementation by elementary quantum gates) way of doing this is the following, which gives us the **quantum (binary) query:**

$$Q_f : |i\rangle \, |j\rangle \, |k\rangle \to |i\rangle \, |j \oplus f(i)\rangle \, |k\rangle,$$

where $\oplus$ denotes addition modulo 2. This type of query is used in Grover's algorithm (to get information about the function f, see the introduction). The binary quantum query model was also studied intensively from the complexity point of view, see Beals, Buhrmann, Cleve, Mosca, and de Wolf [3].

Now we consider the general case, that is, of functions f on general domains D with values in $\mathbb{R}$, as we encounter them in numerical analysis.

A **quantum query** is given by $(m, m_1, m_2, \tau, \beta)$, where $m, m_1, m_2 \in \mathbb{N}$, $m_1 + m_2 \leq m$,

$$\tau : \{0, \ldots, 2^{m_1} - 1\} \to D$$

and

$$\beta : \mathbb{R} \to \{0, \ldots, 2^{m_2} - 1\}$$

are any mappings. The quantum query $Q_f : H_m \to H_m$ is defined as follows, where it is convenient to consider H_m as being represented as $H_m = H_{m_1} \otimes H_{m_2} \otimes H_{m-m_1-m_2}$.

$$Q_f : |i\rangle \, |j\rangle \, |k\rangle \to |i\rangle \, |j \oplus \beta(f(\tau(i)))\rangle \, |k\rangle,$$

with $\oplus$ standing for addition modulo 2^{m_2}. The role of τ is to map indices i to nodes $\tau(i) \in D$, while β encodes the real number $f(\tau(i))$ as a binary integer $\beta(f(\tau(i)))$. This type of query was introduced in [12].

For β one could take, for example,

$$\beta(x) = \begin{cases} 0 & \text{if } x < a \\ \left\lfloor 2^{m_2} \frac{x-a}{b-a} \right\rfloor & \text{if } a \leq x < b \\ 2^{m_2} - 1 & \text{if } x \geq b \end{cases}$$

if we have to deal with functions taking values in $[a,b]$.

A **quantum algorithm** A_n (for the approximate solution of a problem of the type (4.1)) is given by

$$(m, m_1, m_2, \tau, \beta, U_0, \ldots, U_n, i_0, \varphi)$$

where $(m, m_1, m_2, \tau, \beta)$ is a tuple defining a quantum query Q_f as explained above, $U_0, \ldots, U_n$ are unitary operators on H_m, $0 \leq i_0 < 2^m$ is any number (describing the starting state), and $\varphi : \{0, \ldots, 2^m - 1\} \to G$ is any mapping (which produces the final output of the algorithm from the measurement in a classical computation). The computation acts as follows.

Quantum model of computation:

starting state:

$$|i_0\rangle \in H_m \text{ (a classical state)}$$

computation:

$$|i_0\rangle \to U_0 |i_0\rangle \to Q_f U_0 |i_0\rangle \to U_1 Q_f U_0 |i_0\rangle \to \ldots$$

$$\to U_n Q_f U_{n-1} \ldots Q_f U_1 Q_f U_0 |i_0\rangle =: |\xi\rangle$$

measurement:

$$|\xi\rangle = \sum_{i=0}^{2^m-1} \alpha_i |i\rangle \to |i\rangle \text{ with probability } |\alpha_i|^2$$

output:

$$|i\rangle \to \varphi(i) =: A_n(f) \in G$$

We refer to A_n as a quantum algorithm with n queries, the role of the queries being the same as the role of the function values in the classical settings explained in §4.2. So the Q_f (and only these) provide information about the input function f, while the unitaries $U_0, \ldots, U_n$ stand for the compositions of the quantum gates applied between queries to process the obtained information. Observe that $A_n(f)$ is a random variable, with values in the normed space G (which is always $\mathbb{R}$ for our purposes of studying integration). Therefore, the error of A_n at input $f \in F$ is defined in the probabilistic way:

$$e(S, A_n, f) = \inf \{\varepsilon : \Pr\{\|S(f) - A_n(f)\|_G \leq \varepsilon\} \geq 3/4\}.$$

Note that we specified the error probability to be not greater than $1/4$. This special choice is not essential, the number can be replaced by any other number strictly between 0 an $1/2$. The reason is that by repeating an algorithm k times and computing the median of the results, the error probability can be reduced to 2^{-ck} for some $c > 0$ not depending on k.

The error of A_n over the whole class F is

$$e(S, A_n, F) = \sup_{f \in F} e(S, A_n, f).$$

The crucial quantity for complexity analysis is the quantum n-th minimal error

$$e_n^{\mathrm{q}}(S, F) = \inf_{A_n} e(S, A_n, F)$$

that is, the minimal error reachable among all possible quantum algorithms that use not more than n quantum queries. This quantity neglects all combinatory cost (number of gates, classical operations) as it is customary in query-based complexity analysis. However, in deriving matching upper and lower bounds, the proof of the upper ones usually contains a concrete algorithm whose total cost can be counted and, for all the situations studied here, except for Theorem 4.6, turn out to be of the same order (at least up to logarithms) as the number of queries. We therefore just state the asymptotic estimate of $e_n^{\mathrm{q}}(S, F)$, keeping in mind these comments.

4.5 Mean computation and integration

The following result is due to Brassard, Høyer, Mosca, and Tapp [5, 4] (upper bound) and Nayak and Wu [23] (lower bound).

Theorem 4.1 *There is a constant $0 < c < 1$ such that for all n and N, $n < cN$,*

$$e_n^{\mathrm{q}}(S_N, \mathcal{B}(L_\infty^N)) \asymp n^{-1}.$$

The estimate of Theorem 4.1 says two things: First, there is a quantum algorithm which requires not more than n quantum queries and computes the mean for all sequences in $\mathcal{B}(L_\infty^N)$ with error not larger than $c_1 n^{-1}$, $c_1 > 0$ a constant not depending on n and N. (Note that we often use the same symbol for possibly different constants.) Second, it says that no algorithm that uses not more than n quantum queries (and maybe even an unlimited number of gates!) can have error less than $c_2 n^{-1}$, with $c_2 > 0$ another independent of n and N constant.

Let us say some words about the methods behind Theorem 4.1. The counting algorithm of Brassard, Høyer, Mosca, and Tapp [4], originally designed for computing the mean of $\{0, 1\}$-valued sequences, is easily adapted to $[-1, 1]$-valued sequences. This algorithm provides the upper estimate and is based on two fundamental techniques of quantum computing – the technique of Shor of using the quantum version of the discrete Fourier transform for estimating eigenvalues of certain unitary operators, and the Grover iterate, a crucial ingredient of Grover's algorithm. The counting algorithm can be implemented using $\mathcal{O}(n \log^2 n)$ elementary quantum gates (compare the comment at the end of the previous section).

The lower bounds come from the polynomial method [3], which states that the success probability of a quantum algorithm is a certain polynomial of degree at most the number of queries. Starting from that, Nayak and Wu [23] use classical facts from approximation theory: the Bernstein and Markov inequality for polynomials.

Now we want to compare Theorem 4.1 to the known result for the classical deterministic setting. It is clear that the algorithm of direct computation of the mean has error 0. However, it needs all the N function values. What we are particularly interested in (also in view of its applications in high dimensional integration) is a small n and a huge N. Here deterministic algorithms fail completely. We have, for each constant $0 < c < 1$

$$e_n^{\text{det}}(S_N, \mathcal{B}(L_\infty^N)) \asymp 1 \quad (n < cN),$$

that is, no deterministic algorithm using essentially less than N queries can have an error essentially better than the trivial one. In the classical randomized setting we have

$$e_n^{\text{ran}}(S_N, \mathcal{B}(L_\infty^N)) \asymp n^{-1/2}.$$

We see that randomized classical algorithms reach non-trivial error for

n essentially less than N. However, we also see the speedup of quantum algorithms. It is a quadratic one, like in Grover's search algorithm.

Now we present the first result about integration.

Theorem 4.2 (Novak [26]) *Let* $r \in \mathbb{N}_0$, $d \in \mathbb{N}$ *and* $0 < s \leq 1$. *Then*

$$e_n^{\mathrm{q}}(I_d, \mathcal{B}(F_d^{r,s})) \asymp n^{-\frac{r+s}{d}-1}.$$

Novak uses Theorem 4.1 and tools from information-based complexity theory. Moreover, a major ingredient in the proof is a technique from the field of Monte Carlo algorithms – a quantum analog of separation of the main part.

It is instructive to compare this to the classical settings: In the classical deterministic case,

$$e_n^{\mathrm{det}}(I_d, \mathcal{B}(F_d^{r,s})) \asymp n^{-\frac{r+s}{d}}.$$

Look at this statement for d huge, or, in other words, $(r+s)/d$ small. Then the best convergence rate of deterministic classical algorithms is practically negligible, while the quantum rate is smaller than n^{-1}. Let us consider this from the point of view of the number of queries needed to reach error $\varepsilon > 0$. It is readily calculated from the above, that in the classical deterministic setting, we need at least $\Omega((1/\varepsilon)^{d/(r+s)})$ queries. Consequently, the exponent of the cost is proportional to the dimension d. The respective number of quantum queries is $\Omega((1/\varepsilon)^{d/(r+s+d)})$, so this exponent is always smaller than 1. In this sense we can say that quantum algorithms can provide a polynomial speedup of arbitrarily large degree over classical deterministic algorithms. In the classical randomized setting we have

$$e_n^{\mathrm{ran}}(I_d, \mathcal{B}(F_d^{r,s})) \asymp n^{-\frac{r+s}{d}-1/2}.$$

Looking also at this from the point of view of high d, we see that quantum algorithms reach again essentially a quadratic speedup over classical randomized algorithms. (As already mentioned, the statements for the classical settings are known results from information-based complexity theory, and we refer to the sources [31, 25, 11].)

After these results have been obtained, another interesting question arose: The above are results for the L_∞-norm. What about weaker norms like the L_2-norm, that is, what happens for $f \in \mathcal{B}(L_2^N)$ and $f \in \mathcal{B}(W_2^r)$, or more generally $f \in \mathcal{B}(L_p^N)$, $f \in \mathcal{B}(W_p^r)$? We know that the L_2-case is the most important one for Monte Carlo algorithms, and that they preserve the rate of the L_∞-case even for the larger L_2-classes. Could it be that for $2 \leq p < \infty$, quantum algorithms loose gradually, and eventually, for $p = 2$, Monte Carlo algorithms turn out to be as good as quantum algorithms? This problem was solved in [12]. It turns out that there are quantum algorithms which also preserve the favorable rate of the L_∞-case (up to logarithmic factors, at least), and so, the speedup remains also for the L_2-case.

Theorem 4.3 (Heinrich [12]) *Let $1 \leq p < \infty$. There is a constant $c > 0$ such that for all n and N, with $n < cN$,*

$$e_n^q(S_N, \mathcal{B}(L_p^N)) \asymp n^{-1} \quad \text{if } 2 < p < \infty$$
$$e_n^q(S_N, \mathcal{B}(L_p^N)) \asymp_{\log} n^{-2+2/p} \quad \text{if } 1 \leq p \leq 2 \text{ and } n < \sqrt{N}.$$

We use the notation $\asymp_{\log}$ to indicate that the bounds are sharp up to logarithmic factors, that is, the upper and lower bound may contain differing terms of the form $c \log^\alpha n \log^\beta N$ where c, α, β may depend on the problem parameters p (and d and r later on) but do not depend on n and N.

For comparison, in the classical deterministic setting, we have

$$e_n^{\det}(S_N, \mathcal{B}(L_p^N)) \asymp 1,$$

and in the classical randomized setting

$$e_n^{\mathrm{ran}}(S_N, \mathcal{B}(L_p^N)) \asymp \begin{cases} n^{-1/2} & \text{if } 2 \leq p < \infty \\ n^{-1+1/p} & \text{if } 1 \leq p < 2. \end{cases}$$

The new quantum algorithms are based on a suitable multilevel splitting of sequences from L_p^N, distributing queries over levels, and combining the decay of the integral over the levels and precise error estimates for counting. The proof of the lower bound combines techniques of information-based complexity theory with the approach of Nayak and Wu [23].

Theorem 4.3 shows that so far the case of p-summable sequences for $1 \leq p < 2$ and $\sqrt{N} \leq n < N$ was left open. In fact, the upper bound holds true also for that range. So is it sharp? (Since we are interested in huge N and moderate n, this problem was at first glance a rather academic one.) It was a certain surprise that in this case a further improvement of the quantum algorithms of Theorem 4.3 is possible, and moreover, the result turned out to be a crucial ingredient to determine the sharp order in the Sobolev case discussed below.

Theorem 4.4 (Heinrich and Novak [17]) *Let $1 \leq p < 2$. There is a constant $c > 0$ such that for all n and N, with $\sqrt{N} \leq n < cN$,*

$$e_n^q(S_N, \mathcal{B}(L_p^N)) \asymp_{\log} n^{-2/p} N^{2/p-1}.$$

The optimal algorithm contains a new element. It uses Grover's search algorithm to handle a certain portion of the sequence, while the rest of it is taken care by the multilevel type algorithm developed for the proof of Theorem 4.3.

From Theorems 4.3 and 4.4, combined with a new discretization techniques, one can derive optimal quantum integration algorithms for functions from Sobolev spaces. This discretization technique, which is close in spirit to Maiorov's technique from approximation theory [20], allows to reduce the integration problem to a scale of discrete problems – of mean computation in $L_p^{N_l}$ ($l = 1, \ldots, k$) for suitable k and N_l. This technique is useful also in the classical randomized setting, since it can be viewed as a multilevel variance reduction technique for Monte Carlo integration. Applications to the study of Monte Carlo algorithms which use few random bits will be given in [18].

Theorem 4.5 (Heinrich [13]) *Let $1 \leq p < \infty$, $r, d \in \mathbb{N}$, $r/d > 1/p$. Then*

$$e_n^q(I_d, \mathcal{B}(W_{p,d}^r)) \asymp_{\log} n^{-r/d-1}.$$

For comparison, in the classical deterministic setting we have

$$e_n^{\det}(I_d, \mathcal{B}(W_{p,d}^r)) \asymp n^{-r/d}$$

and in the classical randomized setting

$$e_n^{\text{ran}}(I_d, \mathcal{B}(W_{p,d}^r)) \asymp n^{-r/d-1/2} \quad \text{if} \quad 2 \leq p < \infty$$
$$e_n^{\text{ran}}(I_d, \mathcal{B}(W_{p,d}^r)) \asymp n^{-r/d-1+1/p} \quad \text{if} \quad 1 \leq p < 2.$$

Note the following interesting situation in the case $p = 1$. Quantum algorithms are by a factor of about n^{-1} better than classical randomized ones, while the latter yield no improvement over classical deterministic ones.

4.6 Summary and further comments

For a convenient overview we summarize the presented results in a table, including the known results about the classical settings. All rates are sharp (up to possible logarithmic factors, which we suppress, again). We present the $p = 1$ cases separately, because of the interesting relations between the settings.

	e_n^{det}	e_n^{ran}	e_n^{q}
$\mathcal{B}(L_p^N), 2 \leq p \leq \infty$ $n \leq cN$	1	$n^{-1/2}$	n^{-1}
$\mathcal{B}(L_p^N), 1 \leq p < 2$ $n < n^{-1+1/p}\sqrt{N}$,	1		$n^{-2+2/p}$
$\mathcal{B}(L_p^N), 1 < p < 2$ $\sqrt{N} \leq n \leq cN$	1	$n^{-1+1/p}$	$n^{-2/p}N^{2/p-1}$
$\mathcal{B}(L_1^N), n < \sqrt{N}$	1	1	1
$\mathcal{B}(L_1^N), \sqrt{N} \leq n \leq cN$	1	1	$n^{-2}N$
$\mathcal{B}(F_d^{r,s})$	$n^{-(r+s)/d}$	$n^{-(r+s)/d-1/2}$	$n^{-(r+s)/d-1}$
$\mathcal{B}(W_{p,d}^r), 2 \leq p \leq \infty$	$n^{-r/d}$	$n^{-r/d-1/2}$	$n^{-r/d-1}$
$\mathcal{B}(W_{p,d}^r), 1 < p < 2$	$n^{-r/d}$	$n^{-r/d-1+1/p}$	$n^{-r/d-1}$
$\mathcal{B}(W_{1,d}^r)$	$n^{-r/d}$	$n^{-r/d}$	$n^{-r/d-1}$

As argued before, in the case of integration of Hölder functions, the speedup of quantum over deterministic algorithms can be polynomial of arbitrarily high degree. (Similar conclusions hold for the Sobolev case, as long as p is large enough.) The next result shows that there are function classes with even an exponential speedup.

Theorem 4.6 (Heinrich [12]) *Let $\mathcal{E}$ be the set of functions f on $[0, 1]$ such that $\|f\|_{L_\infty} \leq 1$ and for all $k \in \mathbb{N}$ and $s, t \in [0, 1]$, $|s - t| \leq 2^{-k}$ implies $|f(s) - f(t)| \leq k^{-1}$. Then any classical deterministic integration algorithm of error $0 < \varepsilon < 1/32$ has cost at least $\Omega\left(2^{1/(32\varepsilon)}\right)$, while there is a quantum algorithm with error ε using $\mathcal{O}(1/\varepsilon)$ queries and $\mathcal{O}((1/\varepsilon)^2)$ qubits and gates.*

Another example with a similar speedup is path integration, which was considered by Traub and Woźniakowski [32]. Many interesting problems are related to this topic, in particular the study of broader function classes.

Note that in all problems considered so far the output was a single number. This raises an interesting question to be explored: What gain can quantum algorithms bring if the solution is not a number, but a family of numbers, a function. The extreme case is the approximation problem – here S is the identity embedding between some function spaces.

A first approach to approximation was made by Novak, Sloan, and Woźniakowski [27]. They study approximation in huge-dimensional reproducing kernel Hilbert spaces and clarify the conditions of tractability (polynomial dependence on the dimension) in the quantum setting. An interesting problem which is left open is to find matching upper and lower bounds.

Tight bounds for approximation of Sobolev embeddings in the quantum setting were obtained by the author [15].

An interesting numerical problem between integration and approximation is the computation of integrals depending on a (possibly multidimensional) parameter. The classical randomized setting was studied in [19]. The quantum setting was recently considered by Wiegand [33].

References

[1] Abrams, D. S. and Williams, C. P. (1999). Fast quantum algorithms for numerical integrals and stochastic processes (Technical report, http://arXiv.org/abs/quant-ph/9908083).

[2] Aharonov, D. (1998). Quantum computation – a review, *Annual Review of Computational Physics* (World Scientific, volume VI, ed. Dietrich Stauffer). See also http://arXiv.org/abs/quant-ph/9812037.

[3] Beals, R., Buhrman, H., Cleve, R., Mosca, M. and de Wolf, R. (1998). Quantum lower bounds by polynomials, *Proceedings of 39th IEEE FOCS*, 352–361. See also http://arXiv.org/abs/quant-ph/9802049.

[4] Brassard, G., Høyer, P., Mosca, M. and Tapp, A. (2000). Quantum amplitude amplification and estimation (Technical report, http://arXiv.org/abs/quant-ph/0005055).

[5] Brassard, G., Høyer, P. and Tapp, A. (1998). Quantum counting, *Lect. Notes in Comp. Science* **1443**, 820–831. See also http://arXiv.org/abs/quant-ph/9805082.

[6] Deutsch, D. (1985). Quantum theory, the Church-Turing principle and the universal quantum computer, *Proc. R. Soc. Lond., Ser. A* **400**, 97–117.

[7] Ekert, A., Hayden, P. and Inamori, H. (2000). *Basic concepts in quantum computation*. See http://arXiv.org/abs/quant-ph/0011013.

[8] Feynman, R. (1982). Simulating physics with computers, *Int. J. Theor. Phys.* **21**, 467–488.
[9] Grover, L. (1996). A fast quantum mechanical algorithm for database search, *Proc. 28 Annual ACM Symp. on the Theory of Computing* (ACM Press New York), 212–219. See also http://arXiv.org/abs/quant-ph/9605043.
[10] Gruska, J. (1999). *Quantum Computing* (McGraw-Hill, London).
[11] Heinrich, S. (1993). Random approximation in numerical analysis, *Functional Analysis* (Bierstedt, K.D., Pietsch, A., Ruess, W.M. and Vogt, D., editors, Marcel Dekker), 123–171.
[12] Heinrich, S. (2002). Quantum summation with an application to integration, *J. Complexity* **18**, 1–50. See also http://arXiv.org/abs/quant-ph/0105116.
[13] Heinrich, S. (2003). Quantum integration in Sobolev classes, *J. Complexity*, to appear. See also http://arXiv.org/abs/quant-ph/0112153.
[14] Heinrich, S. (2003). From Monte Carlo to Quantum Computation, *Proceedings of the 3rd IMACS Seminar on Monte Carlo Methods MCM2001, Salzburg*, to appear in Mathematics and Computers in Simulation. See also http://arXiv.org/abs/quant-ph/0112152.
[15] Heinrich, S. (2003). Quantum Approximation of Sobolev Embeddings. Paper in preparation.
[16] Heinrich, S. and Novak, E. (2002). Optimal summation and integration by deterministic, randomized, and quantum algorithms, *Monte Carlo and Quasi-Monte Carlo Methods 2000* (Fang, K.-T., Hickernell, F.J. and Niederreiter, H., editors, Springer-Verlag, Berlin), 50–62. See also http://arXiv.org/abs/quant-ph/0105114.
[17] Heinrich, S. and Novak, E. (2003). On a problem in quantum summation, *J. Complexity*, to appear. See also http://arXiv.org/abs/quant-ph/0109038.
[18] Heinrich, S., Novak, E., and Pfeiffer, H. (2003). Paper in preparation.
[19] Heinrich, S. and Sindambiwe, E. (1999). Monte Carlo complexity of parametric integration, *J. Complexity* **15**, 317–341.
[20] Maiorov, V.E. (1975). Discretization of the problem of diameters (in Russian), *Usp. Mat. Nauk 30*, No. 6 (186), 179–180.
[21] Manin, Y.I. (1980). Computable and uncomputable (in Russian), *Sovetskoye Radio, Moscow*.
[22] Manin, Y.I. (1999). Classical computing, quantum computing, and Shor's factoring algorithm. See http://arXiv.org/abs/quant-ph/9903008.
[23] Nayak, A. and Wu, F. (1999). The quantum query complexity of approximating the median and related statistics, *STOC, May 1999*, 384–393. See also http://arXiv.org/abs/quant-ph/9804066.
[24] Nielsen, M.A. and Chuang, I.L. (2000). *Quantum Computation and Quantum Information* (Cambridge University Press).
[25] Novak, E. (1988). Deterministic and Stochastic Error Bounds in Numerical Analysis, *Lecture Notes in Mathematics* **1349** (Springer).
[26] Novak, E. (2001). Quantum complexity of integration, *J. Complexity* **17**, 2–16. See also http://arXiv.org/abs/quant-ph/0008124.

[27] Novak, E., Sloan, I.H. and Woźniakowski, H. (2002). Tractability of Approximation for Weighted Korobov Spaces on Classical and Quantum Computers. See http://arXiv.org/abs/quant-ph/0206023.
[28] Pittenger, A.O. (1999). *Introduction to Quantum Computing Algorithms* (Birkhäuser, Boston).
[29] Shor, P.W. (1994). Algorithms for quantum computation: Discrete logarithms and factoring, *Proceedings of the 35th Annual Symposium on Foundations of Computer Science* (IEEE Computer Society Press, Los Alamitos, CA), 124–134. See also http://arXiv.org/abs/quant-ph/9508027.
[30] Shor, P.W. (2000). Introduction to quantum algorithms. See http://arXiv.org/abs/quant-ph/0005003.
[31] Traub, J.F., Wasilkowski, G.W. and Woźniakowski, H. (1988). *Information-Based Complexity* (Academic Press).
[32] Traub, J.F. and Woźniakowski, H. (2001). Path integration on a quantum computer. See http://arXiv.org/abs/quant-ph/0109113.
[33] Wiegand, C. (2003). Paper in preparation.

5

Straight-line Programs in Polynomial Equation Solving

Teresa Krick

Département de Mathématique
Université de Limoges
France, and
Departamento de Matemática
Universidad de Buenos Aires
Argentina
Email: teresa.krick@unilim.fr, krick@dm.uba.ar

Abstract

Solving symbolically polynomial equation systems when intermediate and final polynomials are represented in the usual dense encoding turns out to be very inefficient: the sizes of the systems one can deal with do not respond to realistic needs. Evaluation representations appeared in this frame a decade ago as a new possibility to treat new families of problems.

We present a survey of the most recent complexity results for different polynomial problems when polynomials are encoded by evaluation (straight-line) programs. We also show surprising mathematical by-products, such as new mathematical invariants and results, that appeared as a consequence of the search of good algorithms.

5.1 Introduction

There are several geometric questions that naturally arise when we are faced to a system of polynomial multivariate equations: do the given equations have at least a common root in an algebraic closure of the base field? If this is so, is there a finite or infinite number of them? What is the dimension of the solution variety? How to describe it in a more tractable manner?

Two major lines have been proposed to answer this kind of questions: numerical analysis which responds with approximate solutions,

---

[0] Partially supported by LACO, UMR CNRS 6090 (France) and UBACyT EX-X198 (Argentina).

and computational algebra with its symbolic procedures giving exact solutions. In this paper we deal with this second aspect, although the evaluation methods we describe tend a natural bridge to the numerical point of view.

Nowadays most usually applied symbolic algorithms rely on rewriting techniques where the input is given by the number of variables, degree bounds and the list of polynomials with (implicitly) all their possible coefficients: this is the case for Gröbner bases computations and for characteristic set descriptions (and also with some minor changes for the more recently considered sparse systems). Unfortunately for the usual case when the degree d of the polynomials is greater than the number n of variables, the size of the input system is typically large, essentially of order d^n, and the degree of the polynomials describing the output can reach d^n as well, which means that writing the output requires at least $(d^n)^n$ symbols, a quantity that is exponential in the size of the input. Moreover, it is a well-known fact that the worst-case complexity of Gröbner bases computations is doubly exponential in n. This behavior prevents us from considering large polynomial equation systems with rewriting techniques.

Evaluation representations began to be strongly considered as an alternative a decade ago. A first and quite naïve motivation of this point of view is that there are polynomials that nobody writes (in dense representation) but everybody computes for specific values, like for example the determinant of a matrix of indeterminates. As another motivation, for the first question raised above — the effective Nullstellensatz — there is a classic example (which gives the well-known lower bound for the degrees of the polynomials arising in a Bézout identity, see §5.3 below) that suggested that there are always for this question Bézout identities composed by polynomials that behave better than expected with respect to evaluation, in the sense that they can be evaluated faster than they should. A careful development of new techniques, that I partially describe here, proved that this intuition was right.

The consideration through evaluation methods (straight-line programs) of the stated geometric questions lets us classify their complexity with respect not only to the usual parameters, given by the number n of variables and the number s and degree d of the input polynomials, but also to less usual parameters like the length L of the straight-line program representation of the input, and the size δ of the underlying linear algebra structure (this parameter, more precisely defined in §5.2.3 and §5.2.5 below, is nowadays called the *geometric degree of the input polynomial system*). It is shown that all considered geometric questions behave polynomially with respect to these parameters: more precisely there are probabilistic algorithms and straight-line program representations for the output polynomials whose complexity and lengths are polynomial in s, n, d, δ and L.

Here is an example of such a result, for the case of a zero-dimensional variety, which represents a core result in this philosophy; see §5.4.1 below.

Theorem 5.1 *Let $f_1, \ldots, f_n \in \mathbb{Q}[x_1, \ldots, x_n]$ be polynomials of degree bounded by d and encoded by straight-line programs of length L, which define a zero-dimensional variety $Z \subset \mathbb{C}^n$. Set δ for the geometric degree of the input polynomial system.*

Then there is a bounded probability algorithm which computes (slp's for) a simple and tractable presentation of Z (a geometric resolution) within complexity $(nd\delta L)^{\mathcal{O}(1)}$.

As the parameters L and δ are in the worst case (and also in a random case) equal to sd^n (for $d \geq n$) and d^n respectively, but may be in some specific cases polynomial in n, s and d, the result can be read as giving an exponential bound in the worst case but a polynomial bound for certain subfamilies of input polynomials.

Also, another consequence of this result is that when the input is codified in the dense representation and its size is measured by sd^n (for $d \geq n$), the length of the straight-line program representation of the output is polynomial in this quantity instead of being exponential as it happens with its dense representation. Since from a straight-line program one clearly (but not rapidly) recovers a dense representation through interpolation, the result implies that the exponential behavior of the complexity of these questions (when considering them classically) is all contained in the final interpolation: there is no exponentiality needed before.

Another research line suggested by this classification is related to the Bézout number: it is usual to associate to a family of s polynomials of degrees $d_1, \ldots, d_s$ in n variables such that $d_1 \geq \cdots \geq d_s$, the Bézout number $D := d_1 \cdots d_{\min\{n,s\}}$. The main property of this Bézout number is that it bounds the geometric degree of the variety defined by the input polynomials. However a precise definition of such a Bézout number D should depend intimately on the representation of the input polynomials: for polynomials of degree d encoded in dense representation, $D := d^n$ seems to be a natural choice, while for sparse polynomials with support in $\mathcal{A}$, $D := \text{Vol}(\mathcal{A})$ seems to be the right notion of Bézout number, as this quantity also controls the degree of the variety.

This digression is motivated by the following crucial observation: in the computation of the resultant, the length of the input L together with the associated Bézout number D and the number of variables n controls the complexity. In the case of dense representation of the input and $d \geq 2$, the typical length L equals $\mathcal{O}(nd^n)$ and $D = d^n$ while for the sparse representation, we have $L \geq 1$ and $D = \text{Vol}(\mathcal{A})$. In both cases, the complexity of computing the resultant is $(nD)^{\mathcal{O}(1)}L$. The optimal complexity estimate should in fact be linear in D as well, although it is not clear what is the exact dependence on n: in the linear case, that is for $n+1$ dense linear forms, $L = \mathcal{O}(n^2)$ and $D = 1$ hold and the resultant equals the determinant, which is conjectured — but still not proved — to be computable in $\mathcal{O}(n^2)$. In an even more general framework, the conjecture is that the computation of (a slp representation of) any geometric object associated to a family of polynomials in n variables represented in a given encoding, with associated Bézout number D and associated length of the input L, should be linear in both D and L, and (possibly) quadratic in n. Here the associated Bézout number D could be the geometric degree of the input polynomial system.

A final comment on the contents of this paper: I only treat here results concerning *upper bounds* for the sequential complexity of *geometric* questions. I do not consider algebraic questions like for instance the ideal membership problem since their complexity is usually accepted to behave essentially worse. Also, all bounds depend on the size of the underlying linear algebra structure which is in the worst case of order d^n independently from the fact $d \geq n$ or not. In case $d = 2$ and n arbitrary, the size of the input is of order n^2 instead of 2^n while our algorithms are in the generic case polynomial in 2^n. A completely different analysis and novel approach are needed to deal with this case. Finally, concerning *lower bounds* — a task of a different order of complexity as everybody knows — there is a deep research actually going on: we refer to [12] and the references given there for an overview of the most recent and striking results on the matter.

This paper is voluntarily written in a non-technical style: for each subject I tried to priorize ideas and natural developments over precise definitions, proofs or full generality of results: references where these can be found are always given. The paper is divided into six sections. §1 concerns with a very quick and intuitive introduction on data structures and algorithms, priorizing properties of the straight-line program encodings with respect to other encodings, and also with some preliminaries needed for the sequel. §2 presents the effective Nullstellensatz as a motivation of the spirit of the paper. It contains a presentation of classic upper and lower (degree and arithmetic) bounds and a discussion on the utility of evaluation methods with a succinct idea of an algorithm. In particular it shows how a good evaluation method combined with a deep and non-trivial arithmetic analysis yield optimal bounds for the arithmetic Nullstellensatz. §3 concentrates on zero-dimensional varieties, presenting geometric resolutions (shape lemmas descriptions) and Chow forms and comparing both characterizations of these varieties. §4 gives the generalizations of these notions to equidimensional varieties of arbitrary dimension, and introduces Newton's method to lift the information on a good zero-dimensional fibre to the corresponding positive-dimensional component. §5 shows an outline of a general algorithm which describes each equidimensional component of a variety from a set description. This algorithm is mainly the result of many other algorithms performing related tasks that were developed and improved during the last 5 years and are somewhat discussed during the whole paper. Finally §6 gives a couple of applications that are interesting on their own, even in a more classical frame.

Many of the ideas and algorithms surveyed in this paper are implemented in a MAGMA package, called *Kronecker*, developed by Grégoire Lecerf [49].

5.2 Preliminaries

5.2.1 Data structures

The objects we deal with are polynomials in n variables with coefficients in a field k of characteristic zero. That is

$$f = \sum_\alpha a_\alpha x^\alpha \quad \text{with} \quad a_\alpha \in k,$$

where $\alpha := (\alpha_1, \ldots, \alpha_n) \in \mathbb{N}_0^n$ and $x^\alpha := x_1^{\alpha_1} \cdots x_n^{\alpha_n}$.

We insist on the fact that the characteristic of the base field k is zero, for some of the techniques and results we present do not apply for positive characteristic. The notation $\mathbb{A}^n$ always refers to $\mathbb{A}^n(\overline{k})$, where $\overline{k}$ is an algebraic closure of k, unless otherwise specified.

The usual *dense encoding* for representing such a polynomial f is given by an a priori bound d for the degree of f and an array of the $\binom{d+n}{d} = \binom{d+n}{n}$ coefficients a_α (zero coefficients as well as non-zero ones) in a pre-established order.

In opposition the *sparse encoding* only represents the non-zero coefficients by means of couples $(\alpha; a_\alpha)$ indicating the exponent α corresponding to a non-zero coefficient a_α. (Another classic way of defining sparsity is fixing the Newton polytope allowed, that is the convex hull of the exponents corresponding to non-zero coefficients, we only consider it here in the applications.)

In this paper we deal with a third way of representing a polynomial f, which is called the *straight-line program encoding* (*slp* for short). The idea of using slp as short encodings of special families of polynomials goes back to the seventies, when it appeared in questions concerning the probabilistic testing of polynomial identities. The first applications to computer algebra dealt with the elimination of one variable problems [35, 41, 42]. Later there were extended to multivariate elimination problems by Marc Giusti, Joos Heintz and their collaborators, in works that are partly reviewed here.

There are many different slp approaches. We refer to [8] for the standard definition or to [35, 45] for other models. We only describe here the simplest one, in a non-rigorous manner that we hope is enough for the readability of this paper:

Definition 5.1 *Given a polynomial $f \in k[x_1, \ldots, x_n]$, a slp encoding of f is an evaluation circuit γ for f, where the only operations allowed belong to $\{+, -, \cdot\}$ (no divisions) and the constants $a \in k$ can be used freely.*

More precisely: $\gamma = (\gamma_{1-n}, \ldots, \gamma_0, \gamma_1, \ldots, \gamma_L)$ where $f = \gamma_L$, $\gamma_{1-n} := x_1, \ldots, \gamma_0 := x_n$ and for $k > 0$, γ_k is of one of the following forms:

$$\gamma_k = a * \gamma_j \quad \text{or} \quad \gamma_k = \gamma_i * \gamma_j \quad \text{where } a \in k, \ * \in \{+, -, \cdot\} \text{ and } i, j < k.$$

For example, the dense encoding of the polynomial $f = x^{2^d}$ (in 1 variable) is $(1, 0, \ldots, 0)$ for the decreasing order of monomials, its sparse encoding equals $(2^d; 1)$ and a straight-line program encoding is for instance given by the following slp γ:

$$\gamma_0 = x, \ \gamma_1 = \gamma_0 \cdot \gamma_0 = x^2, \ \gamma_2 = \gamma_1 \cdot \gamma_1 = x^{2^2}, \ldots, \gamma_d = \gamma_{d-1} \cdot \gamma_{d-1} = x^{2^d}.$$

We specify now the *lengths* associated to these encodings: here we assume that each constant of the field k has length 1. (In many concrete situations the input polynomials have integer or rational coefficients and thus a more realistic measure of the input is given by taking also into account a bound for the maximum bit length of every integer allowed to appear.) Thus the dense encoding of a polynomial f of degree bounded by d like above has length $\binom{d+n}{d} = \mathcal{O}(d^n)$ (at least if $d \geq n$ as it is usually the case), while the sparse encoding has length $(n+1)N$ where N is a bound for the number of non-zero coefficients of f. Finally the length of a slp γ like above is defined as $L(\gamma) = L$ (note that $\gamma_{1-n}, \ldots, \gamma_0$ are added to the list only to handle with the variables and therefore have no cost), and the length $L(f)$ of f is the minimum of the lengths of γ for γ a slp encoding f.

Coming back to the example, the length of the dense encoding of x^{2^d} is $2^d + 1$, the length of its sparse encoding is 2 while the length of its slp encoding is bounded by d since we exhibited a slp γ for f such that $L(\gamma) = d$. However, note that for $(x+y)^{2^d}$ (in 2 variables) one can produce immediately a slp γ' of length $d+1$ defining $\gamma'_1 := x+y$ and then squaring like in γ, while both the dense and the sparse encodings have length $\binom{2^d}{2} = \mathcal{O}(2^{2d})$. This observation is an example of the following crucial fact:

Remark 5.1 *Straight-line programs behave well under linear changes of variables (while sparsity does not).*

Now let us compare dense encoding and slp encoding lengths. Every polynomial has a standard slp encoding given essentially by its dense encoding:

Remark 5.2 *Let $f \in k[x_1, \ldots, x_n]$ be a polynomial of degree d, then*

$$L(f) \leq 3\binom{d+n}{d}.$$

Proof One shows inductively that for any $r \in \mathbb{N}$, there is a slp of length bounded by $\binom{n+r}{r}$ whose intermediate results are all monomials x^α with $|\alpha| \leq r$ (once one has a list of all the monomials of degree bounded by $r-1$, each one of the $\binom{n+r-1}{r}$ homogeneous monomials of degree r is simply obtained from one of the list multiplying by a single variable). Finally we multiply all monomials of f by their coefficients and add them up, that is we add $2\binom{d+n}{d}$ instructions to obtain a slp encoding for f. □

Also, it is clear that a sparse polynomial has a "short" slp (if one knows in advance a bound for the degree): Let $f \in k[x_1, \ldots, x_n]$, $\deg f \leq d$, be a polynomial with at most N non-zero coefficients, then $L(f) \leq Nd + N - 1$.

Reciprocally, if a polynomial $f \in k[x_1, \ldots, x_n]$ is represented by a slp of length L and *a bound for its degree d is known*, its dense encoding is trivially obtained within $d^{\mathcal{O}(n)} L(f)$ operations, simply interpolating in a grid of $(d+1)^n$ points. Of course this is not very satisfactory since we loose the possible benefit we had of having a short slp for f. However, it is important to notice that polynomials with short slp's are very rare. This is an important classification fact:

Fix a bound d for the degree of the polynomials. In the same way that sparse polynomials (we mean polynomials with at least one prescribed zero coefficient) belong to the union of closed hyperplanes of the set of all polynomials, polynomials with slp's essentially shorter that the length given by the standard dense encoding belong to a closed hypersurface of the set of all polynomials; see [35] or [34] Th. 3.2:

Proposition 5.1 *For every n, d and $c \in \mathbb{N}$, there exists a hypersurface $\mathcal{H} \subset \mathbb{A}^{\binom{n+d}{d}}$ such that*

$$\{f \in k[x_1, \ldots, x_n], \deg f \leq d \text{ and } L(f) \leq (nd)^c\}$$
$$\Rightarrow f = \sum a_\alpha x^\alpha \text{ with } (a_\alpha)_\alpha \in \mathcal{H}.$$

Roughly speaking this fact says that a random polynomial of degree d takes essentially as much time to be evaluated than its whole number of (zero and non-zero) monomials. Polynomials like in the statement of Proposition 5.1 are very special, and are nowadays called *smart polynomials*. We will show that quite amazingly the polynomials that naturally appear when dealing with geometric questions related to polynomial equations are smart.

A bad feature of slp encodings is that two different slp's may encode the same polynomial, or more simply a slp can encode the zero polynomial, without our noticing. Of course, even if we know the degree of f, evaluating in a grid of $(d+1)^n$ points is forbidden for too expensive. There is in this line a remarkable result due to Heintz and Schnorr ([34] Th. 4.4) that shows that there exist test grids (*correct test sequences*) whose cardinality depend polynomially on the slp length of the polynomial:

Lemma 5.1 *Let $\mathcal{F} := \{f \in k[x_1, \ldots, x_n] : \deg f \leq d, L(f) \leq L\}$. There exists in any big enough set of k (whose size depends polynomially on d and L) a subset $\mathcal{A}$ with $\#\mathcal{A} = (nL)^{\mathcal{O}(1)}$ such that:*

$$\forall f \in \mathcal{F}, \ f(a) = 0 \ \forall \ a \in \mathcal{A}^n \ \Rightarrow \ f = 0.$$

This is an existential result and nobody knows until now how to exhibit economically such correct test sequences. For the design of probabilistic algorithms one can replace it by the Zippel-Schwartz zero test ([68, 59]):

Lemma 5.2 *Let $\mathcal{A} \subset k$ be a finite set. For any $f \in k[x_1, \ldots, x_n]$, $f \neq 0$, the probability that a randomly chosen $a \in \mathcal{A}^n$ annihilates f verifies*

$$\Pr(f(a) = 0) \leq \frac{\deg f}{\#\mathcal{A}}.$$

5.2.2 Algorithms

The formalization of our algorithms is given by the Blum-Shub-Smale machine over k with the restriction that the only branches allowed are comparisons to zero. Roughly speaking the algorithm is a finite sequence of instructions performed on the input, where each instruction can be an arithmetic operation $(+, -, \cdot)$ on elements of k or a comparison to zero and a selection of how to continue depending on the result of the comparison. We refer to [5] Ch. 3 and 4. The special feature here is that most of the algorithms we refer to compute as their output slp encodings instead of lists of coefficients (dense or sparse encodings). For many of them, the input is also encoded by slp's; see [37] Sec. 1.2 for a more formal presentation.

In some cases we refer to *bounded probability algorithms*, algorithms with special nodes that flip coins (these nodes randomly choose the following instruction between two possible ones with probability 1/2 for each of them ([5] Sec. 17.1 and [37] Sec. 1.2) so that the error probability of the result of the algorithm is bounded by 1/4. In our setting probability is introduced by choosing a random element a with equidistributed probability in a set $\{0, 1, \ldots, N-1\}^n$ where a certain polynomial f of known degree will be specialized in order to apply Zippel-Schwartz zero-test.

The complexity or time of the algorithm is equal to the number of arithmetic operations performed (each arithmetic operation on k has unit cost), comparisons, selections and flipping coins can be considered with no cost since if they are meaningful their number is bounded by the number of operations. Again this model can be adequate to more realistic needs, e.g. counting bit operations in an integer setting.

5.2.3 Parameters

We adopt the following parameters to measure an input polynomial system $f_1, \ldots, f_s \in k[x_1, \ldots, x_n]$: the number of variables n, a bound for the degrees d, the number of polynomials s, the maximum length L of slp's computing $f_1, \ldots, f_s$ and also a parameter δ which measures the maximum dimension of the underlying linear algebra structures. This new parameter appeared naturally during the search of good algorithms with slp encodings, and is mentioned for the first time in [25]. It is associated to the input polynomials and is called the *geometric degree of the input polynomial system*. It is in the worst case bounded by the Bézout number d^n although it can be substantially smaller.

In case $s \leq n+1$ and $f_1, \ldots, f_s$ is a reduced weak regular sequence, that is, for $1 \leq i \leq s-1$, f_{i+1} is not a zero-divisor modulo the ideal $(f_1, \ldots, f_i)$ which is a radical ideal (this implies in particular that for $1 \leq i \leq s$, the variety $V(f_1, \ldots, f_i)$ is pure of codimension i), the parameter δ is defined as

$$\delta := \max_{1 \leq i \leq s} \deg(V(f_1, \ldots, f_i))$$

where deg denotes the usual geometric affine degree of the variety.

In case the input polynomials $f_1, \ldots, f_s$ do not define a reduced weak regular sequence, we perturb them performing a sufficiently generic scalar combination: for a *good choice* of $a_1, \ldots, a_{n+1} \in k^s$ (the meaning of good choice is explained in §5.2.5 below), we define the polynomials $\tilde{f}_1, \ldots, \tilde{f}_{n+1}$ as

$$\tilde{f}_1 := a_{11}f_1 + \cdots + a_{1s}f_s, \ldots, \tilde{f}_{n+1} := a_{n+1 1}f_1 + \cdots + a_{n+1 s}f_s,$$

and we define a geometric degree δ (associated to $a := (a_1, \ldots, a_{n+1})$ of the input polynomial system as

$$\delta(a) := \max_{1 \leq i \leq n+1} \deg(V(\tilde{f}_1, \ldots, \tilde{f}_i)).$$

The definition given here is a simplified version of the many different definitions of geometric degree of the input polynomial system that appear in different papers, each time adapted to their context. In particular we only choose *a* geometric degree depending of the good choice a, which is enough for our purpose, and skip the definition of *the* geometric degree which is an intrinsic quantity that does not depend on the choice of a.

5.2.4 Basic linear algebra ingredients

Our algorithms rely on the possibility of performing the usual linear algebra operations by means of algorithms behaving well with slp's. For instance the computation of (slp's for) the coefficients of the characteristic polynomial of a $D \times D$ matrix, as well as the computation of its adjoint and its determinant, can be done within $\mathcal{O}(D^4)$ arithmetic operations with no divisions and no branches [3].

Another useful fact is that a slp of length L for the computation of a polynomial $f \in k[x_1, \ldots, x_n]$ of degree bounded by d produces easily slp's of length $\mathcal{O}(d^2 L)$ for the homogeneous components of any given degree of f; see [45] Lem. 13 and [8] Lem. 21.25.

Also, there is a classic division free algorithm known as Strassen's Vermeidung von Divisionen [64] which computes a slp for the quotient of two polynomials provided it is a polynomial. More precisely

Proposition 5.2 *Let $f, g \in k[x_1, \ldots, x_n]$ be polynomials encoded by slp's of length L such that $f(0) = 1$. Assume that f divides g in $k[x_1, \ldots, x_n]$ and that $\deg g/f \le d$. Then there is an algorithm which computes a slp for g/f within complexity $\mathcal{O}(d^2(d + L))$.*

The idea is simply to use that

$$f^{-1} = \frac{1}{1 - (1 - f)} = \sum_{k \ge 0} (1 - f)^k$$

and to truncate all operations and the result at order d. This algorithm is easily adapted to more general situations when $f(a) \ne 0$ for $a \in k^n$, or when $f \ne 0$ and one looks probabilistically for $a \in k^n$ such that $f(a) \ne 0$.

Finally there is a bounded probability algorithm to compute the greatest common divisor of two multivariate polynomials encoded by slp's [41].

5.2.5 Input preparation

Given $f_1, \ldots, f_s \in k[x_1, \ldots, x_n]$ which define an arbitrary variety $V := V(f_1, \ldots, f_s) \subset \mathbb{A}^n$, as many authors do we replace the original input system by taking a linear combination of the polynomials and a change of variables, in order to attain the good underlying linear algebra structure we discussed partly in §5.2.3.

- In case $f_1, \ldots, f_s$ are not (known to be) a reduced regular sequence we replace them by $\tilde{f}_1, \ldots, \tilde{f}_{n+1}$ as explained in §5.2.3, for a choice of $a = (a_1, \ldots, a_{n+1}) \in k^{(n+1) \times s}$ such that:

  - $V(\tilde{f}_1, \ldots, \tilde{f}_{n+1}) = V$.

- For $0 \leq r \leq n-1$, if $V(\tilde{f}_1, \ldots, \tilde{f}_{n-r}) \neq V$, then $I_r := (\tilde{f}_1, \ldots, \tilde{f}_{n-r})$ is a radical ideal of dimension r outside V (that is every primary component $\mathcal{Q}$ of I_r such that $V(\mathcal{Q}) \not\subset V$ is prime of dimension r).

These conditions imply that if a minimal equidimensional decomposition of V is given by

$$V = V_0 \cup \cdots \cup V_{n-1}$$

where for $0 \leq r \leq n-1$, V_r is either empty or equidimensional of dimension r, then

$$V(I_r) = V'_r \cup V_r \cup \cdots \cup V_{n-1}$$

where V'_r is either empty or an equidimensional variety of dimension r (that contains in particular all the components of lower dimension of V).

In case the original variety $V := V(f_1, \ldots, f_s)$ is empty, the perturbed polynomials $\tilde{f}_1, \ldots, \tilde{f}_{n+1}$ verify that for a certain $t \leq n$, $(\tilde{f}_1, \ldots, \tilde{f}_t)$ is a reduced regular sequence and $V(\tilde{f}_1, \ldots, \tilde{f}_{t+1}) = \emptyset$.

An important fact is that Bertini's theorem insures that for a generic choice of such a matrix a, the desired conditions are always attained; see for instance [1] Sec. 4, [27] Sec. 3.2, [58] Prop. 18 and proof of Th. 19. Moreover, the coefficients of the matrices a giving bad choices belong to a hypersurface of degree bounded by $4(d+1)^{2n}$; see [50] Lem. 1 and 2 or [46] Prop. 4.3 and Cor. 4.4. This enables us to apply Zippel-Schwartz zero test.

- We replace the variables $x_1, \ldots, x_n$ by new variables $y_k = b_{k1}x_1 + \cdots b_{kn}x_n$, $1 \leq k \leq n$, such that for $0 \leq r \leq n-1$, the variables $y_1, \ldots, y_r$ are in Noether normal position with respect to the equidimensional component W_r of $V(I_{n-r})$ of dimension r, that is, the morphism $\pi : W_r \to \mathbb{A}^r$, $y \mapsto (y_1, \ldots, y_r)$ is finite of degree $\deg W_r$.

In fact we look for a more technical condition (see Assumption 5.1 below) which implies this Noether position one. Again the important fact is that a generic choice of the new variables insures the desired conditions. Moreover, the coefficients of the matrices b giving bad choices belong to a hypersurface of degree bounded by $n(n-1)d^{2n}$ ([46] Prop. 4.5), which enables us to apply Zippel-Schwartz zero test again.

5.3 The Nullstellensatz

This section discusses results on the effective Nullstellensatz that motivate the spirit of this survey paper. It also presents some complexity aspects in more detail.

The (weak) Nullstellensatz states (for a field k with algebraic closure $\overline{k}$):

Let $f_1, \ldots, f_s$ be polynomials in $k[x_1, \ldots, x_n]$. The equation system

$$f_1(x) = 0, \ldots, f_s(x) = 0$$

has no solution in $\overline{k}^n$ if and only if there exist $g_1, \ldots, g_s \in k[x_1, \ldots, x_n]$ satisfying the Bézout identity

$$1 = g_1 f_1 + \cdots + g_s f_s. \tag{5.1}$$

Upper bounds

Bounds for the degrees of polynomials g_i's satisfying Identity (5.1) immediately yield a linear system of equations. Showing such bounds is what is nowadays called *Effective Nullstellensätze*.

In 1926, Hermann [36] (see also [31], [54]) proved that in case Identity (5.1) holds, there exist $g_1, \ldots, g_s \in k[x_1, \ldots, x_n]$ with $\deg g_i f_i \leq 2(2d)^{2^{n-1}}$. After a conjecture of Keller and Gröbner, this estimate was dramatically improved by Brownawell [7] to $\deg g_i f_i \leq n^2 d^n + nd$ in case char$(k) = 0$, while Caniglia, Galligo and Heintz [9] showed that $\deg g_i f_i \leq d^{n^2}$ holds in the general case.

These results were then independently refined by Kollár [43] and by Fitchas and Galligo [19] to

$$\deg g_i f_i \leq \max\{3, d\}^n, \tag{5.2}$$

which is optimal in case $d \geq 3$. For $d = 2$, Sombra [62] showed that the bound $\deg g_i f_i \leq 2^{n+1}$ holds.

A lower bound

We turn now to a lower bound estimate. The following well-known example due to Masser and Philippon yields a lower bound for any general degree estimate. Set

$$f_1 := x_1^d, \; f_2 := x_1 - x_2^d, \; \ldots, \; f_{n-1} := x_{n-2} - x_{n-1}^d, \; f_n := 1 - x_{n-1} x_n^{d-1}$$

for any positive integers n and d. These are polynomials of degree d in n variables. Let $g_1, \ldots, g_n \in \mathbb{Q}[x_1, \ldots, x_n]$ be polynomials satisfying

Bézout identity (5.1). Specializing it at

$$x_1 := t^{(d-1)d^{n-2}}, x_2 := t^{(d-1)d^{n-3}}, \ldots, x_{n-1} := t^{d-1}, x_n := 1/t \quad \text{for } t \neq 0$$

one obtains

$$1 = g_1(t^{(d-1)d^{n-2}}, \ldots, t^{d-1}, 1/t) \, t^{(d-1)d^{n-1}}$$

which implies that $\deg_{x_n} g_1 \geq (d-1)d^{n-1}$.

In fact here is a Bézout identity with optimal degrees for these polynomials:

$$1 = x_n^{(d-1)d^{n-1}} x_1^d - x_n^{(d-1)d^{n-1}} (x_1^d - (x_2^d)^d) - \cdots$$
$$- x_n^{(d-1)d^{n-1}} (x_{n-2}^{d^{n-2}} - (x_{n-1}^d)^{d^{n-2}}) + (1 - (x_{n-1}x_n^{d-1})^{d^{n-1}})$$

i.e. $g_1 = x_n^{(d-1)d^{n-1}}, g_2 = -g_1(x_1^{d-1} + \cdots + (x_2^d)^{d-1}), \ldots, g_n = 1 + x_{n-1}x_n^{d-1} + \cdots + (x_{n-1}x_n^{d-1})^{d^{n-1}-1}$.

This example immediately shows that the dense encoding of any output $g_1, \ldots, g_n$ has length at least $\binom{d^n - d^{n-1} + n}{n}$, which is exponential in the length $\binom{d+n}{n}$ of the dense encoding of the input. Moreover, a slight perturbation of this example — replacing x_n by a linear combination of the variables — destroys all the sparsity of the output.

However in this case there is at least one choice of *smart* polynomials since a coarse computation shows that $L(g_1) \leq n(d-1)$ and $L(g_i) \leq (n+5i)(d-1)$. Here we have used the identity:

$$x^{d^i-1} + x^{d^i-2}y + \cdots + y^{d^i-1}$$
$$= (x^{d-1} + x^{d-2}y + \cdots + y^{d-1}) \cdots$$
$$(x^{d^{i-1}(d-1)} + x^{d^{i-1}(d-2)}y + \cdots + y^{d^{d-1}(d-1)}).$$

Arithmetic bounds

Now let us consider the arithmetic aspects of the Nullstellensatz, that is when the input polynomials have integer coefficients (or more generally coefficients in a number ring). In the case of integer coefficients, the Nullstellensatz takes the following form:

Let $f_1, \ldots, f_s \in \mathbb{Z}[x_1, \ldots, x_n]$ be polynomials such that the equation system

$$f_1(x) = 0, \ldots, f_s(x) = 0$$

has no solution in $\mathbb{C}^n$. Then there exist $a \in \mathbb{Z} \setminus \{0\}$ and $g_1, \ldots, g_s \in \mathbb{Z}[x_1, \ldots, x_n]$ satisfying the Bézout identity

$$a = g_1 f_1 + \cdots + g_s f_s.$$

Let $h(f)$ denote the *height* of an arbitrary polynomial $f \in \mathbb{Z}[x_1,\ldots,x_n]$, defined as the logarithm of the maximum absolute value of its coefficients, and for the sequel set $h := \max_i h(f_i)$. A slight modification of Masser and Philippon example yields the lower bound $h(a) \geq d^n h$.

On the other hand the bound (5.2) reduces Bézout identity (5.1) to a system of $\mathbb{Q}$-linear equations, which by application of Cramer rule gives an estimate for the height of a and the polynomials g_i of type $s\, d^{n^2}\, (h + \log s + d)$.

It was soon conjectured that the optimal height bound should be closer to the mentioned lower bound than to this trivial upper bound.

Philippon [55] obtained the first sharp estimate for the denominator a in Bézout identity: $\deg g_i \leq (n+2)\, d^n$, $h(a) \leq \kappa(n)\, d^n (h+d)$, where $\kappa(n)$ depends exponentially on n. Then the first essential progress on height estimates for all the polynomials g_i was achieved by Berenstein and Yger who, from 1991 to 1999 [1, 2], obtained $\deg g_i \leq n(2n+1)d^n$, $h(a), h(g_i) \leq \lambda(n)\, d^{4n+3}\, (h + \log s + d \log d)$, where $\lambda(n)$ is a (non-explicit) constant which depends exponentially on n. Their proof relies on the previous work of Philippon and on techniques from complex analysis. Using the algebraic techniques described in Paragraph "Idea of an algorithm" below, the author and Pardo [44, 45] obtained the same kind of estimates though less precisely. In 1998 Sombra convinced us that the techniques were better than the obtained results and that what was lacking was a deeper height analysis. This lead to the nowadays best and essentially optimal arithmetic bound [46] stated in Paragraph "Computational results" below.

Idea of an algorithm

Since 1993, Heintz, Giusti and their collaborators initiated a strong current area of research on computational issues related to the Nullstellensatz [27, 20, 45, 25, 24, 30]. The fact that the polynomials g_i's satisfying Bézout identity in Masser and Philippon counterexample were smart did not seem to be a coincidence. They searched for arguments and tools behaving well under specializations in order to generalize this fact. A good algorithmic answer is given by the application of the duality theory for Gorenstein algebras to this setting. We refer to E. Kunz [48] Appendix F for a complete mathematical presentation of the duality theory.

The initial spirit of the algorithm is quite simple. It works by successive divisions:

$$1 \in (f_1, \ldots, f_s) \iff f_s \text{ is invertible} \quad (\text{mod } (f_1, \ldots f_{s-1}))$$

and more generally, once $g_s, \ldots, g_{i+1}$ are determined

$$1 - g_s f_s - \cdots - g_{i+1} f_{i+1} \in (f_1, \ldots, f_i) \iff$$
$$\exists g_i : 1 - g_s f_s - \cdots - g_{i+1} f_{i+1} \equiv g_i f_i \pmod{(f_1, \ldots f_{i-1})}.$$

To illustrate the algorithm assume now that $(f_1, \ldots, f_s)$ define the empty variety, that $s = n + 1$ and that $I_r := (f_1, \ldots, f_{n-r})$ is an ideal of dimension r for $0 \le r \le n - 1$.

Thus $I_0 := (f_1, \ldots, f_n)$ is a zero-dimensional ideal, and the first step is straight-forward: $B := k[x_1, \ldots, x_n]/I_0$ is a finite-dimensional k-vector space. Therefore an inverse g_{n+1} for f_{n+1} in B can be obtained for instance using the characteristic polynomial χ_{n+1} of (the multiplication by) f_{n+1} in B and Cayley-Hamilton theorem: $\chi_{n+1}(t) = t^D + c_{D-1} t^{D-1} + \cdots + c_0$ (with $c_0 \in k^*$ since f_{n+1} is invertible) implies that we can define

$$g_{n+1} := -\frac{1}{c_0}(f_{n+1}^{D-1} + c_{D-1} f_{n+1}^{D-2} + \cdots + c_1). \tag{5.3}$$

For the second recursion step, even if one can mimic the finite-dimensional vector space argument, in the best case the frame is a finite-rank free module $B := k[x_1, \ldots, x_n]/I_1$ over $A := k[x_1]$, and the argument above fails since here $c_0 \in k[x_1]$ does not necessarily divide the expression in the numerator of the corresponding formula (5.3).

The trace formula giving the duality theory of Gorenstein algebras is the tool which enables us to generalize the previous argument to the case when we are not in a finite vector space frame. It performs effective divisions modulo complete intersection ideals. It was introduced in the context of the effective Nullstellensatz in [20], and then refined in [58, 45, 47, 24, 30]. The latest optimal results for the arithmetic aspects when the base ring is a number ring are obtained in [46].

Here we describe only the basic aspects of the theory we need to sketch the proof.

Let $I_r = (f_1, \ldots, f_{n-r}) \subset k[x_1, \ldots, x_n]$ be a reduced complete intersection ideal (of dimension r), such that $B := k[x_1, \ldots, x_n]/I_r$ is a finite-rank free module over $A := k[x_1, \ldots, x_r]$.

The dual A-module $B^* := \mathrm{Hom}_A(B, A)$ can be seen as a B-module with scalar multiplication defined by $f \cdot \tau(g) := \tau(fg)$ for $f, g \in B$ and $\tau \in B^*$. It happens to be a free B-module of rank 1. Any of its generators is called a *trace* of B. There is a canonical trace σ associated to the complete intersection I_i, and particular polynomials

$a_m, b_m \in k[x_1, \ldots, x_n]$ verifying the following *trace formula*:

$$\forall g \in k[x_1, \ldots, x_n], \quad g \equiv \sum_m \sigma(g a_m) b_m \quad (\mathrm{mod}\ I_r).$$

The canonical trace σ is related to the usual trace Tr of B/A by the equality $\mathrm{Tr}\,(g) \equiv \sigma(Jg) \pmod{I_r}$ where J is the Jacobian determinant of the complete intersection I_r with respect to the variables $x_{i+1}, \ldots, x_n$.

Now we are able to describe — at least theoretically — the second recursion step. All steps follow the same pattern.

Let $I_1 = (f_1, \ldots, f_{n-1})$, $B = k[x_1, \ldots, x_n]/I_1$ and $A := k[x_1]$ be in the hypothesis of the duality theory. Let $\chi_n(t) := t^D + c_{D-1} t^{D-1} + \cdots + c_0$ be the characteristic polynomial of f_n in B/A. Observe that $c_0 \in A \setminus \{0\}$ since f_n is not a zero-divisor modulo I_1. We define

$$f_n^* := f_n^{D-1} + c_{D-1} f_n^{D-2} + \cdots + c_1,$$

$$g_n := -\frac{1}{c_0} \sum_m \sigma(f_n^*(1 - g_{n+1} f_{n+1}) a_m) b_m.$$

Fact: g_n belongs to $k[x_1, \ldots, x_n]$ (i.e. c_0 divides the numerator) and $g_n f_n \equiv 1 - g_{n+1} f_{n+1} \pmod{I_1}$.

Proof.-

- In fact $c_0 \mid \sigma(f_n^*(1 - g_{n+1} f_{n+1}) a_m)$ in $A = k[x_1]$ for every m:

  Since by hypothesis there exists $q \in k[x_1, \ldots, x_n]$ such that $1 - g_{n+1} f_{n+1} \equiv q f_n \pmod{I_1}$ and on the other hand $f_n^* f_n \equiv -c_0 \pmod{I_1}$, we infer that $f_n^*(1 - g_{n+1} f_{n+1}) \equiv -c_0 q \pmod{I_1}$. Therefore

  $$\sigma(f_n^*(1 - g_{n+1} f_{n+1}) a_m) = \sigma(-c_0 q\, a_m) = -c_0 \sigma(q\, a_m)$$

  since σ is a A-morphism and $c_0 \in A$.

- By the trace formula, $-c_0\, g_n \equiv f_n^*(1 - g_{n+1} f_{n+1}) \equiv -c_0 q \pmod{I_1}$. Thus $c_0\, g_n\, f_n \equiv c_0\,(1 - g_{n+1} f_{n+1}) \pmod{I_1}$. Since c_0 is not a zero-divisor modulo I_1 we conclude that $g_n f_n \equiv 1 - g_{n+1} f_{n+1} \pmod{I_1}$.

We finally observe that the relationship between this trace σ and the canonical trace Tr allows to replace in the computations σ that one does not know by Tr which is computable as a coefficient of the characteristic polynomial. The polynomials a_m, b_m are also easily computable.

Computational results

The foundational paper of this computational current of research on the Nullstellensatz is the one of Giusti, Heintz and Sabia [27] followed by [20]:

5. Straight-line Programs

Theorem 5.2 *Let $f_1, \ldots, f_s \in k[x_1, \ldots, x_n]$ be polynomials of degree bounded by d. Then there is a bounded probability algorithm of size $s^{\mathcal{O}(1)} d^{\mathcal{O}(n)}$ which decides whether the ideal $(f_1, \ldots, f_s)$ is trivial or not, and in case it is, produces slp's of the same length for polynomials $g_1, \ldots, g_s \in k[x_1, \ldots, x_n]$ satisfying the Bézout identity $1 = g_1 f_1 + \cdots + g_s f_s$. The degree of these polynomials was first bounded by $d^{\mathcal{O}(n^2)}$ [27] Sec. 2, Th. and then by $d^{\mathcal{O}(n)}$ [20] Th. 2.*

This result follows after an input preparation of the kind of the one described in §5.2.5 in order to place the input in the hypothesis of the duality theory, and a recursive application of the division procedure. The canonical trace is computed as a coefficient of the characteristic polynomial of the multiplication map in B/A. A suitable basis of the natural zero-dimensional vector space associated to B/A is obtained reducing to the two variables case.

Later on, the input polynomials were no more considered in their dense encoding: the complexity bounds are now given in terms of the lengths of the slp encoding and of the geometric degree of the input polynomials [25] Th. 20, [24] Th. 4, Th. 21, that is what is called *intrinsic Nullstellensatz*:

Theorem 5.3 *Let $f_1, \ldots, f_s \in k[x_1, \ldots, x_n]$ be polynomials of degree bounded by d given by slp's of length bounded by L with no common zeroes in $\overline{k}^n$. Let δ be a geometric degree of the input equation system. Then there is a bounded probability algorithm of size $(sn d\delta L)^{\mathcal{O}(1)}$ which produces slp's of the same length for polynomials $g_1, \ldots, g_s \in k[x_1, \ldots, x_n]$ satisfying the Bézout identity $1 = g_1 f_1 + \cdots + g_s f_s$. The degree of these polynomials are bounded by $n^2 d\delta$.*

The proof of this theorem is based on the techniques of [27] in what concerns the recursive divisions. The dependence on δ is due to the precise results on the degrees of [58] and [47, 61] where bounds in terms of δ were first computed. In order to obtain a final bound depending polynomially on L (and not on d^n) the authors introduced a formal version of Newton's method which produces with good complexity good bases of the complete intersection ideals recursively considered. This method is essential in all further developments and will be introduced in a simple frame in §5.5.1.

Now let us add the arithmetic aspects of the Nullstellensatz.

The duality technique introduced above also yields arithmetic bounds. That was done in [44, 45]: the slp produces an integer a and polynomials g_i's such that $\deg g_i \leq (n\,d)^{c\,n}$, $h(a), h(g_i) \leq (n\,d)^{c\,n}(h + \log s + d)$, where c is a universal constant. Then an arithmetic analogue of the intrinsic Nullstellensatz was obtained in [30, 29]. To this aim the authors introduced the notion of *height of a polynomial system*, the arithmetic analogue of the geometric degree of the system. In [46] these results are generalized and brought to an optimal form. As a consequence of this intrinsic statement, a sparse version is obtained, recently improved by Sombra in [63].

More precisely, the main result of [46] in its simplest form is the following:

Theorem 5.4 *Let $f_1, \ldots, f_s \in \mathbb{Z}[x_1, \ldots, x_n]$ be polynomials without common zeros in $\mathbb{C}^n$. Set $d := \max_i \deg f_i$ and $h := \max_i h(f_i)$.*
Then there exist $a \in \mathbb{Z} \setminus \{0\}$ and $g_1, \ldots, g_s \in \mathbb{Z}[x_1, \ldots, x_n]$ such that

- $a = g_1 f_1 + \cdots + g_s f_s$,
- $\deg g_i \leq 4 n d^n$,
- $h(a), h(g_i) \leq 4 n (n+1) d^n (h + \log s + (n+7) \log(n+1) d)$.

The proof of this arithmetic Nullstellensatz also relies on the trace formula. However there is another key ingredient which is the notion of local height of a variety defined over a number field K introduced there:

For $V \subset \mathbb{A}^n(\overline{\mathbb{Q}})$ an equidimensional affine variety defined over K and for an absolute value v over K, the *local height* $h_v(V)$ of V at v is defined — inspired by results of Philippon — as a Mahler measure of a suitable normalized Chow form of V. This definition is consistent with the Falting's height $h(V)$ of V, namely:

$$h(V) = \frac{1}{[K : \mathbb{Q}]} \sum_{v \in M_K} N_v \, h_v(V),$$

where M_K denotes the set of canonical absolute values of K, and N_v the local degree of K at v. Then the authors obtained estimations of the local height of the trace and the norm of a polynomial $f \in K[x_1, \ldots, x_n]$ with respect to an integral extension $K[\mathbb{A}^r] \hookrightarrow K[V]$. There are also local analogues of many of the global results of Bost, Gillet and Soulé [6] and Philippon [56].

5.4 Zero-dimensional varieties

We devote this section to the description of a zero-dimensional variety by means of two different presentations: a classic description that we call here, following [24] Sec. 2.1, a geometric resolution of the variety (also known as a shape lemma presentation or a rational univariate representation), and its Chow form (also known as the u-resultant when associated to a system of equations). We compare both approaches.

For the whole section, $Z \subset \mathbb{A}^n$ denotes a 0-dimensional variety (that is a finite variety) of cardinality D.

5.4.1 Geometric resolutions

Geometric resolutions were first introduced in the works of Kronecker and König in the last years of the XIX century. Nowadays they are widely used in computer algebra. We refer to [23] for a complete historical account.

A *geometric resolution* of Z consists of an affine linear form $\ell(x) = u_0 + u_1 x_1 + \cdots + u_n x_n \in k[x_1, \ldots, x_n]$ and of polynomials $q \in k[t]$ and $w = (w_1, \ldots, w_n) \in k[t]^n$ (where t is a new single variable) such that:

- The affine linear form ℓ is a *primitive element* of Z, that is $\ell(\xi) \neq \ell(\xi')$ for all $\xi \neq \xi'$ in Z.
- The polynomial q is monic of degree D and $q(\ell(\xi)) = 0$ for all $\xi \in Z$; that is,
$$q(t) = \prod_{\xi \in Z} (t - \ell(\xi))$$
is the minimal polynomial of ℓ over Z.
- For $1 \leq i \leq n$, $\deg w_i < D$ and
$$Z = \{(w_1(\ell(\xi)), \ldots, w_n(\ell(\xi))); \ \xi \in Z\}$$
$$= \{(w_1(\tau), \ldots, w_n(\tau)); \ \tau \in \overline{k} \ / \ q(\tau) = 0\};$$
that is, w parametrizes Z by the zeroes of q.

Observe that the minimal polynomial q and the parametrization p are uniquely determined by the variety Z and the affine linear form ℓ. We say that (q, w) is *the geometric resolution of Z associated to ℓ*.

The existence of such a geometric resolution of Z (at least with coefficients in $\overline{k}$) is simple to show:

Let $\ell(x) = 0$ be any projection hyperplane that separates the zeroes of Z (any generic enough hyperplane will do), and define $q(t) = \prod_{\xi \in Z} (t - \ell(\xi))$. Thus q is a polynomial of degree $D := \#Z$ that vanishes on Z. Now for $1 \leq i \leq n$, let $w_i(t)$ be the unique polynomial of degree strictly bounded by D which verifies that $w_i(\ell(\xi)) = \xi_i$ for every

$\xi = (\xi_1, \ldots, \xi_n) \in Z$. Then the polynomial $x_i - w_i(\ell(x))$ also vanishes on Z and it is easy to show that in fact

$$Z = V\left(q(\ell(x)), x_1 - w_1(\ell(x)), \ldots, x_n - w_n(\ell(x))\right).$$

The computation of a geometric resolution

The algorithm that we comment here has its beginning in [26] and the ideas were then refined in [25] with the introduction of Newton's method and in [24] where the use of computable companion matrices replaced theoretical algebraic roots. Further improvements were then developed independently in [51, 28] and [33] where a significant speed-up is obtained by a technique called deforestation.

Here, in order to simplify the presentation we assume that the zero-dimensional variety Z is given as the zero set of a reduced regular sequence $f_1, \ldots, f_n$ in $k[x_1, \ldots, x_n]$.

Theorem 5.5 ([24] Th. 19) *Let $f_1, \ldots, f_n \in k[x_1, \ldots, x_n]$ be polynomials of degree bounded by d and encoded by slp's of length L. Assume that the polynomials are a reduced regular sequence and set δ for a geometric degree of the input polynomial system.*

Then there is a bounded probability algorithm which computes (slp's for) a separating linear form ℓ and a geometric resolution (q, v) of Z associated to ℓ within complexity $(nd\delta L)^{\mathcal{O}(1)}$.

The algorithm has n recursive steps: it adds one equation at a time. For simplicity one assumes that $x_1, \ldots, x_n$ are in Noether normal position with respect to the ideals $(f_1, \ldots, f_i)$, $1 \leq i \leq n$.

The i-th step computes from a geometric resolution of the zero-dimensional variety

$$Z_i := V(f_1, \ldots, f_i) \subset \mathbb{A}^i(\overline{k(x_1, \ldots, x_{n-i})})$$

a geometric resolution of the zero-dimensional variety

$$Z_{i+1} := V(f_1, \ldots, f_{i+1}) \subset \mathbb{A}^{i+1}(\overline{k(x_1, \ldots, x_{n-i-1})}).$$

The first input is given by the geometric resolution $(q(t) := f_1(x_1, \ldots, x_{n-1}, t), w_1(t) := t)$ of $Z_1 := V(f_1)$ associated to the separating linear form $\ell := x_n$, and the last $(n-1)$ step computes a geometric resolution of the zero dimensional variety $Z_n = Z$.

The crucial point here is that the input of step $i+1$ cannot be simply the output of step i, where the natural length of this output would be

$L_{i+1} = (nd\delta_i)^{\mathcal{O}(1)} L_i$ (where δ_i is the size of the underlying linear algebra at step i), since in that case the recursion would yield an output length

$$L_n = (nd)^{\mathcal{O}(n)} (\delta_0 \cdots \delta_{n-1})^{\mathcal{O}(1)} L = (nd\delta)^{\mathcal{O}(n)} L$$

which does not represent any improvement with respect to other known algorithms.

The alternative was for the first time addressed in [26] where the authors dealt with the necessity of a compression of the input data at each recursive step that enabled them to add L_i instead of multiplying it : $L_{i+1} = (nd\delta_i)^{\mathcal{O}(1)} L + L_i$. Another principal breakthrough of this paper is that it adapted the concept of geometric resolution to a positive dimension context, rediscovering Kronecker's approach.

The general form of an algorithm like this one is considered in more detail for the computation of Chow forms in §5.6, after the introduction of Newton's method and the use of companion matrices in some simple cases.

5.4.2 Chow forms

Set $L(U, x) := U_0 + U_1 x_1 + \cdots + U_n x_n$ for a generic (affine) linear form where $U := (U_0, \ldots, U_n)$ denotes a new group of variables. Typically a specialized linear form $\ell(x) := L(u, x)$ does not meet any of the points of Z unless $u \in \mathbb{A}^n$ is a root of the following polynomial

$$\mathcal{C}h_Z(U) = \prod_{\xi \in Z} L(U, \xi).$$

This polynomial is called the (normalized) *Chow form* of Z. It happens to be a homogeneous polynomial in $k[U]$ of degree D. We refer to [60] Sec. I.6.5 for the proof of this fact.

Thus, the main feature of the Chow form is that for any $u \in k^{n+1}$,

$$\mathcal{C}h_Z(u) = 0 \iff Z \cap \{L(u, x) = 0\} \neq \emptyset.$$

Chow forms $\to$ geometric resolutions

A Chow form gives straightforward a "generic" geometric resolution and hence, by specialization, families of geometric resolutions. We describe here the procedure, essentially due to Kronecker:

The polynomial $\mathcal{P}_Z(U, t) \in k[U, t]$ (where t is a single variable like before) defined by

$$\mathcal{P}_Z(U, t) := (-1)^D \mathcal{C}h_Z(U_0 - t, U_1, \ldots, U_n) = \prod_{\xi \in Z} (t - L(U, \xi))$$

verifies that $P(U, x) := \mathcal{P}_Z(U, L(U, x)) = \sum_\alpha a_\alpha(x) U^\alpha$ vanishes clearly on every $\xi \in Z$. Thus, for every α, $a_\alpha(\xi) = 0$ for every $\xi \in Z$, which implies in particular that $\frac{\partial P(U,x)}{\partial U_i}$ also vanishes on every $\xi \in Z$.

Now for $1 \leq i \leq n$,

$$\frac{\partial P}{\partial U_i}(U, x) = \frac{\partial \mathcal{P}_Z}{\partial U_i}(U, L(U, x)) + \frac{\partial \mathcal{P}_Z}{\partial t}(U, L(U, x)) x_i$$

implies that for every $\xi \in Z$,

$$\frac{\partial \mathcal{P}_Z}{\partial U_i}(U, L(U, \xi)) + \frac{\partial \mathcal{P}_Z}{\partial t}(U, L(U, \xi)) \xi_i = 0.$$

This last equality means that for every $u \in \overline{k}^{n+1}$ such that both $\ell(x) := L(u, x)$ verifies that $\ell(\xi) \neq \ell(\xi')$ for all $\xi \neq \xi'$ in Z and $\frac{\partial \mathcal{P}_Z}{\partial t}(u, \ell(\xi)) \neq 0$ for all $\xi \in Z$ (these conditions are fulfilled in a nonempty open Zariski subset of $\overline{k}^{n+1}$), one has that

$$\xi_i = -\frac{\frac{\partial \mathcal{P}_Z}{\partial U_i}(u, \ell(\xi))}{\frac{\partial \mathcal{P}_Z}{\partial t}(u, \ell(\xi))}.$$

A proper geometric resolution of Z associated to ℓ is then given by $q(t) := \mathcal{P}_Z(u, t)$ and the polynomials $w_i(t)$ that one can obtain using the discriminant $\varrho(U)$ of $\mathcal{P}_Z(U, t)$ with respect to t to eliminate the polynomial $\frac{\partial \mathcal{P}_Z}{\partial t}(u, \ell(\xi))$ appearing in the denominator (replacing it by the non-zero constant $\varrho(u)$).

Geometric resolutions $\to$ Chow forms:

Now let us show how to derive the Chow form from a given geometric resolution of Z with respect to a linear form ℓ. This simple and beautiful construction relies on the fact that even if we do not know the coordinates of each zero ξ of Z, a geometric resolution gives the information of the zeroes altogether:

We are looking for

$$\mathcal{C}h_Z(U) = \prod_{\xi \in Z} L(U, \xi) = |\mathrm{Diag}_{\xi \in Z}(L(U, \xi))|,$$

where $|\mathrm{Diag}(\ )|$ denotes the determinant of the diagonal matrix with the entries under the brackets in the diagonal. But the information we have is that of $q(t) = \prod_{\xi \in Z}(t - \ell(\xi))$ whose companion matrix C_q is similar ($\sim$) to $\mathrm{Diag}_{\xi \in Z}(\ell(\xi))$ since $\ell(\xi) \neq \ell(\xi')$ for $\xi \neq \xi'$. For $1 \leq i \leq n$, we also have w_i such that $\xi_i = w_i(\ell(\xi))$. Thus

$$w_i(C_q) \sim w_i(\mathrm{Diag}_{\xi \in Z}(\ell(\xi))) \sim \mathrm{Diag}_{\xi \in Z}(w_i(\ell(\xi))) \sim \mathrm{Diag}_{\xi \in Z}(\xi_i).$$

We infer that
$$L\bigl(U, (\mathrm{Id}, w_1(C_q), \ldots, w_n(C_q))\bigr) \sim \mathrm{Diag}_{\xi \in Z}\bigl(L(U, \xi)\bigr)$$
and we conclude by taking the determinant of the left hand side.

This beautiful application of companion matrices is a crucial tool that was introduced in this context in [24] pp. 285-286 to replace each zero in Z by their "all-together information".

5.5 Equidimensional varieties

A variety is said to be equidimensional if all its irreducible components have the same dimension. We recall that the degree of an equidimensional variety V is defined as the number of points in the intersection of V with a generic linear variety of codimension equal to the dimension of V.

To simplify the presentation, we set $n = r + m$ and we distinguish the variables in two groups: the set of free variables $y = (y_1, \ldots, y_r)$ and the set of dependent variables $x = (x_1, \ldots, x_m)$ of the extension $k[V]$: for that purpose we assume for the whole section that $V \subset \mathbb{A}^n = \mathbb{A}^{r+m}$ is an equidimensional variety of dimension r and degree D, defined by polynomials in $k[y_1, \ldots, y_r, x_1, \ldots, x_m]$, which satisfies the following assumption:

Assumption 5.1 *We assume that* $Z := V \cap \{y_1 = 0, \ldots, y_r = 0\}$ *is a zero-dimensional variety of cardinality* $\#Z = \deg V = D$.

Assumption 5.1 implies that the variables $y_1, \ldots, y_r$ are in Noether normal position with respect to V [46] Lem. 2.14. That means that if we set $A := k[y_1, \ldots, y_r]$, $A \cap I(V) = \{0\}$ holds, and that for $1 \leq i \leq m$, there is an integral dependence equation for x_i over A modulo $I(V) \subset A[x_1, \ldots, x_m]$: there exists a non-zero and monic polynomial $p_i \in A[x_i] \cap I(V)$. We remark that the previous condition is satisfied by any variety under a generic linear change of variables.

5.5.1 Geometric resolutions

We present here the notions of geometric resolution of an equidimensional variety of positive dimension.

Under Assumption 5.1 we can reduce easily to the zero-dimensional case: we invert the variables $y_1, \ldots, y_r$. We set $K := k(y_1, \ldots, y_r)$ for the field of fractions of $A = k[y_1, \ldots, y_r]$ and we consider the following objects:

$$I^e := K[x_1, \ldots, x_m] \cdot I(V) \subset K[x_1, \ldots, x_m]$$
$$V^e := V(I^e) \subset \overline{K}^m,$$

where $I(V)$ is the ideal of V and $V(I^e)$ is the variety defined by I^e. V^e is a zero-dimensional variety of cardinality $D = \deg V$ and a geometric resolution of V is (essentially) given by a geometric resolution of V^e. It does not describe the whole variety V but it describes it outside a given hypersurface. It consists of an affine linear form $\ell = u_0 + u_{r+1}x_1 + \cdots + u_{r+m}x_m \in k[x_1, \ldots, x_m]$ and of polynomials $q \in A[t]$ and $w = (w_1, \ldots, w_m) \in A[t]^m$ such that:

- The affine linear form ℓ is a *primitive element* of V^e, that is $\ell(\xi) \neq \ell(\xi')$ for all $\xi \neq \xi'$ in V^e.
- The polynomial q, of degree D, is the monic minimal polynomial of ℓ with respect to the extension $K \hookrightarrow K[V^e]$. The Noether position assumption guarantees that the coefficients of q belong to A [28] Sec. 3.2.
- For $1 \le i \le m$, $\deg_t w_i < D$ and $\varrho\, x_i = w_i(\ell)$ in $K[V^e]$, where $\varrho \in A$ is the discriminant of q with respect to t. The polynomial w_i also belongs to $A[t]$ for the same reason.

Thus, we infer that

$$V^e = \left\{ \left(\frac{w_1(\ell(\xi))}{\varrho}, \ldots, \frac{w_m(\ell(\xi))}{\varrho} \right) ; \xi \in V^e \right\}$$
$$= \left\{ \left(\frac{w_1(\tau)}{\varrho}, \ldots, \frac{w_m(\tau)}{\varrho} \right) ; \tau \in \overline{K} \,/\, q(\tau) = 0 \right\}.$$

In particular, since q is monic in t, for every $\eta = (\eta_1, \ldots, \eta_r) \in \mathbb{A}^r$ such that $\varrho(\eta) \neq 0$, $\#(V \cap \{y_1 = \eta_1, \ldots, y_r = \eta_r\}) = D$ and the D roots $(\eta, \xi_\eta) \in \mathbb{A}^{r+n}$ are obtained via the D different roots τ_η of $q(\eta, t) = 0$:

$$V \cap \{y_1 = \eta_1, \ldots, y_r = \eta_r\}$$
$$= \left\{ \left(\eta_1, \ldots, \eta_r, \frac{w_1(\eta, \tau_\eta)}{\varrho(\eta)}, \ldots, \frac{w_m(\eta, \tau_\eta)}{\varrho(\eta)} \right) \text{ for } \tau_\eta \text{ s.t. } q(\eta, \tau_\eta) = 0 \right\}.$$

For simplicity of notations, we say that outside the hypersurface $\{\varrho = 0\} \subset \mathbb{A}^{r+m}$ the variety $V \subset \mathbb{A}^{r+m}$ coincides with

$$\left\{ \left(y, \frac{w(y, \tau_y)}{\varrho(y)} \right) \text{ for } \tau_y \text{ s.t. } q(y, \tau_y) = 0 \right\}.$$

We say that (q, w) is the geometric resolution of V associated to ℓ. It gives a simple description of V outside the discriminant variety. There is another equivalent approach, more suitable algorithmically, where the geometric resolution of V is defined outside the variety $\{q' = 0\}$ where $q' = \partial q/\partial t$ instead of outside the discriminant variety (cf. §5.4.2).

Observe that Assumption 5.1 implies that

$$Z = V \cap \{y_1 = 0, \ldots, y_r = 0\}$$
$$= \left\{ \left(0, \frac{w(0, \tau_0)}{\varrho(0)}\right) \text{ for } \tau_0 \text{ s.t. } q(0, \tau_0) = 0 \right\}.$$

Next section is crucial: it shows how the tractable information of the r-dimensional equidimensional variety V is enclosed in the arbitrary input equations plus the information of the zero-dimensional fibre Z. In other terms it shows how to recover a geometric resolution of V from a geometric resolution of Z, lifting the points $(0, \xi_0) \in Z$ to the corresponding $(y, \xi_y) \in V$.

5.5.2 Dimension zero → positive dimension

Let $V \subset \mathbb{A}^{r+m}$ be an equidimensional variety of dimension r satisfying Assumption 5.1 and set as usual $Z := V \cap \{y_1 = 0, \ldots, y_r = 0\}$. Suppose we are given a geometric resolution (q_Z, w_Z) of Z associated to a separating linear form ℓ. How can we derive from it a geometric resolution (q_V, w_V) of V?

The major tool here is the application of Newton's method to lift from an "approximate zero", that is a geometric resolution of the zero-dimensional fibre Z, the geometric resolution of V. Newton's method has been applied in a similar way to recover the exact factorization of multivariate polynomials from factorization algorithms for univariate polynomials by E. Kaltofen in [40]. For polynomial systems it has been previously used by W. Trinks [66] and by F. Winkler [67]. In our specific frame it has been re-introduced by J. Heintz et al. in [25].

First let us recall Hensel's lifting, that is, the algebraic version of Newton's method, in its classic presentation:

Proposition 5.3 *Let p be a prime integer number, $f \in \mathbb{Z}[x]$ and $\xi_0 \in \mathbb{Z}$ such that*

$$f(\xi_0) \equiv 0 \pmod{p}, \quad f'(\xi_0) \not\equiv 0 \pmod{p}.$$

Then, for all $k \in \mathbb{N}$, there exists $\xi_k \in \mathbb{Z}$ such that
$$f(\xi_k) \equiv 0 \pmod{p^{2^k}}, \quad \xi_k \equiv \xi_0 \pmod{p}.$$

The existence (and also uniqueness mod p^{2^k}) of this sequence of integers is given by the recursive application of Newton's operator
$$N_f(x) = x - \frac{f(x)}{f'(x)}$$
to the input approximate zero ξ_0:
$$\text{for all } k \geq 1, \ \xi_k \equiv N_f(\xi_{k-1}) \pmod{p^{2^k}}.$$

Hensel's lifting translates directly to a constructive implicit function theorem, that is the version we use here:

Proposition 5.4 Set $A := k[y_1, \ldots, y_r]$ and $A[x] := A[x_1, \ldots, x_m]$. Let $V \subset \mathbb{A}^{r+m}$ be an equidimensional variety of dimension r defined by a reduced regular sequence $f_1, \ldots, f_m \subset A[x]$, and assume moreover that V satisfies Assumption 5.1.

Set $\mathcal{M} = (y_1, \ldots, y_r) \subset A$ for the maximal ideal associated to 0, $F(y, x) := (f_1(y, x), \ldots, f_m(y, x))$ and $D_x F := (\partial f_i / \partial x_j)_{1 \leq i,j \leq m}$.

Let $\xi_0 \in \mathbb{A}^m$ be such that $F(y, \xi_0) \in \mathcal{M}$ (i.e. $F(0, \xi_0) = 0$, that is $(0, \xi_0) \in Z$) and $|D_x F|(y, \xi_0) \notin \mathcal{M}$. Then the recursive application of
$$N_F(x^t) := x^t - (D_x F(y, x))^{-1} F(y, x)^t$$
initialized at ξ_0 approximates the corresponding fiber root $(y, \xi_y) \in V$ with quadratic precision. That is, if N_F^k denotes the application of k times the Newton operator, $N_F^k(\xi_0)$ tends quadratically to a m-tuple of formal power series $\xi_y \in \overline{k}[[y]]^m$ verifying:

- $F(y, \xi_y) = 0$
- $\xi_y(0) = \xi_0$.

where "tends quadratically" means that $F(y, N_F^k(\xi_0)) \in \mathcal{M}^{2^k}$.

Next section gives an idea of how to derive the polynomial q_V of a geometric resolution of V from a geometric resolution of Z:

Idea of the algorithm

We adopt the abusive notation $N_F^\infty(\xi_0) := \xi_y$.

For the rest of the section, let $V = V(f_1, \ldots, f_m) \subset \mathbb{A}^{r+m}$ be an equidimensional variety of dimension r satisfying Assumption 5.1 which is in the hypothesis of Proposition 5.4. We will recover the polynomial q_V of a geometric resolution of V from a geometric resolution of Z associated to a linear form ℓ, lifting the roots $(0, \xi_0) \in Z$ to their corresponding fiber roots $(y, \xi_y) \in V$ (via the inversion $\overline{k}[[y]] \hookrightarrow \overline{K}$).

Without loss of generality we identify Z with $\{\xi_0 : f_1(0, \xi_0) = \cdots = f_m(0, \xi_0) = 0\} \subset \mathbb{A}^m$.

We know that the total degree of $q_V \in A[t]$ equals D, and we observe that $\ell \in k[x_1, \ldots, x_m]$ is also a separating linear form for V^e.

The information we have is $q_Z(t) = \prod_{\xi_0 \in Z}(t - \ell(\xi_0)) \in k[t]$ and $w := w_Z \in k[t]$ such that for every $\xi_0 \in Z$, $\xi_0 = w(\ell(\xi_0)) = (w_1(\ell(\xi_0)), \ldots, w_1(\ell(\xi_0)))$.

Here is a very informal sketch of how things work:

By Proposition 5.4 we know that for every $\xi_0 \in Z$, $N_F^\infty(\xi_0) = \xi_y \in \overline{k}[[y]]^m$. Thus we are looking for

$$q_V(y, t) = \prod_{(y, \xi_y)} (t - \ell(\xi_y)) = \prod_{(y, \xi_y)} (t - \ell(N_F^\infty(\xi_0))).$$

As we have the a priori bound D for the degree of q_V in the variables y as well, to obtain it *exactly* it is enough to approximate each root ξ_y by a n-tuple of power series up to order D, that is to compute $\lceil \log_2 D \rceil$ iterations of Newton operator on ξ_0 and then to truncate the obtained polynomial at degree D.

Of course we don't know the roots $\xi_0 \in Z$, but we are looking in fact for the characteristic polynomial of the diagonal matrix $\text{Diag}_{\xi_y}(\ell(\xi_y))$, and — as at the end of §5.4 — we have the information of all $\xi_0 \in Z$ together via the companion matrix $C = C_{q_Z}$ of q_Z: for $1 \leq i \leq m$,

$$w_i(C) \sim \text{Diag}_{\xi_0 \in Z}((\xi_0)_i) \implies$$
$$N_F^\infty(w(C)) \sim (\text{Diag}_{\xi_0}((N_F^\infty(\xi_0))_1), \ldots, \text{Diag}_{\xi_0}((N_F^\infty(\xi_0))_m))$$
$$\sim (\text{Diag}_{\xi_y}((\xi_y)_1), \ldots, \text{Diag}_{\xi_y}((\xi_y)_m)) \implies$$
$$\ell(N_F^\infty(w(C))) \sim \text{Diag}_{\xi_y}(\ell(\xi_y)),$$

and we should conclude taking its characteristic polynomial which is exactly the polynomial q_V we are looking for.

Again, as we have the a priori bound D for the degree of q_V, it is enough to compute all approximations up to order D and then to truncate the obtained characteristic polynomial at degree D as well.

Let us conclude with a word on the computational aspects:

We set $k := \lceil \log_2 D \rceil$ and we compute formally polynomials $g_1, \ldots, g_m$ and h corresponding to the numerators and a single denominator of the k-th iteration of Newton operator on a m-tuple of indeterminate variables $(x_1, \ldots, x_m)$. We apply them on the matrices $w_1(C), \ldots, w_m(C)$.

Using Cayley-Hamilton theorem we invert the matrix $h(w(C))$ modulo its determinant in $k[y]$, which is invertible as a power series. We approximate the inverse by a formal power series truncated at order D. All remaining computations are truncated at order D. The details of this procedure can be found for instance in [32] Proof of Th. 2, which deals with a generalization of what is presented here.

Generalizations of Newton-Hensel symbolic lifting where the strong hypothesis made here are weakened, allowing multiplicities, are being deeply studied by Grégoire Lecerf ([51, 28, 52, 53] and work in progress).

5.5.3 Chow forms

For a detailed mathematical account of Chow forms we refer to [60] Sec. I.6.5, [21] Ch. 3, [15], and to [46] Sec. 1.2.2 for the specific normalization introduced here.

Let $V \subset \mathbb{A}^n = \mathbb{A}^{r+m}$ be as before an equidimensional variety of dimension r and degree D satisfying Assumption 5.1 (although unnecessarily heavy here, we decided to keep for the sake of coherence the notation $y = (y_1, \ldots, y_r)$ for the free variables and $x = (x_1, \ldots, x_m)$ for the dependent ones, and we set $n := r + m$).

Generically a linear variety of codimension $r+1$ does not meet V. Like in the zero-dimensional case, the condition on the coefficients of these linear varieties to meet V is given by a polynomial called a Chow form of V. We formalize that: For $i = 0, \ldots, r$, let $U_i = (U_{i0}, U_{i1}, \ldots, U_{in})$ be a group of $n+1$ variables and set $U := (U_0, \ldots, U_r)$, and $L(U_i, (y, x)) := U_{i0} + U_{i1} y_1 + \cdots + U_{in} x_m$ for the associated generic linear form in the variables (y, x).

Define

$$\Phi_V = \{(u_0, \ldots, u_r; \xi), \xi \in V, L(u_0, \xi) = 0, \ldots, L(u_r, \xi) = 0\}$$
$$\subset (\mathbb{A}^{n+1})^{r+1} \times \mathbb{A}^n,$$

and denote by $\pi : (\mathbb{A}^{n+1})^{r+1} \times \mathbb{A}^n \to (\mathbb{A}^{n+1})^{r+1}$; $(u, \xi) \mapsto u$ the canonical projection. Then the Zariski closure of the image of Φ_V, $\overline{\pi(\Phi_V)} \subset (\mathbb{A}^{n+1})^{r+1}$, is a closed hypersurface [60] p. 66. We define the *Chow form of V* as any squarefree defining equation $\mathcal{F}_V \in k[U_0, \ldots, U_r]$ of $\overline{\pi(\Phi_V)}$.

The main feature of the Chow form is that for every $u_0, \ldots, u_r \in \mathbb{P}^n$,

$$\mathcal{F}_V(u_0, \ldots, u_r) = 0 \Leftrightarrow \overline{V} \cap \{L^h(u_0, (y, x)) = 0\} \cap \cdots$$
$$\cap \{L^h(u_r, (y, x)) = 0\} \neq \emptyset.$$

Here $L^h(U_i,(y,x)) = U_{i0}y_0 + U_{i1}y_1 + \cdots + U_{in}x_m$ stands for the homogeneization of L and $\overline{V} \subset \mathbb{P}^n$ for the projective closure of V.

A Chow form $\mathcal{F}_V$ is a multihomogeneous polynomial of degree D in each group of variables U_i ($0 \le i \le r$). The projective closure $\overline{V} \subset \mathbb{P}^n$ is uniquely determined by a Chow form of V ([60] p. 66). Moreover, it is possible to derive equations for the variety $\overline{V}$ from a Chow form of V ([21] Ch. 3, Cor. 2.6). In case V is irreducible, $\mathcal{F}_V$ is a irreducible polynomial and, in the general case, a Chow form of V of the equidimensional variety V is the product of the Chow forms of its irreducible components.

Observe that the Chow form of an equidimensional variety is uniquely determined up to a scalar factor. Here we follow [46] Sec. 1.2.2 and define *the (normalized) Chow form* Ch_V by fixing the choice of this scalar factor through the condition

$$Ch_V(e_0, \ldots, e_r) = 1,$$

where e_i denotes the $(i+1)$-vector of the canonical basis of k^{n+1}. That is the coefficient of the monomial $U_{00}^D \cdots U_{rr}^D$ (the fact that this coefficient is not zero follows from Assumption 5.1, which says in particular that $\overline{V} \cap \{y_0 \ne 0\} \cap \{y_1 = 0\} \cap \cdots \cap \{y_r = 0\} \ne \emptyset$).

Chow forms $\to$ geometric resolutions:
The procedure described in §5.4.1 is generalizable to any dimension: We define

$$\mathcal{P}(U_0,t,y) := (-1)^{\deg V} Ch_V((U_{00}-t, U_{01}, \ldots, U_{0n}); e_1 - y_1 e_0;$$
$$\ldots; e_r - y_r e_0)$$

For every $\xi = (y, \xi_y) \in V$, we observe that

$$\mathcal{P}(U_0, L(U_0, \xi), y_1, \ldots, y_r) = 0$$

since

$$V \cap \{L(U_0,(y,x)) = L(U_0,(y,\xi_y))\} \cap \{y_1 = y_1\} \cap \cdots \cap \{y_r = y_r\} \ne \emptyset.$$

Thus $\mathcal{P}(U_0, L(U_0, \xi), y_1, \ldots, y_r)$ vanishes on V^e. Moreover $\deg_t \mathcal{P} = \deg V$ since $\deg_{U_0} Ch_V = \deg V$ and for a generic u_0, $\ell(y,x) := L(u_0,(y,x))$ separates the zeroes of V^e. Finally $\mathcal{P}$ is monic in t since it can be shown that its leading coefficient is independent from y, and $Ch_V(-e_0, e_1, \ldots, e_r) = (-1)^{\deg V}$. Then we conclude like in §5.4.1.

Geometric resolutions $\to$ Chow forms
If we proceed exactly like in the pure zero-dimensional case we obtain $Ch_{V^e}(U_0) \in K[U_0]$ in a single set of variables U_0 and with extraneous

coefficients depending on the free variables y (cf. the Chow form of [50] Sec. 3.3).

The first polynomial although probabilistic algorithm to compute the Chow form of an equidimensional variety satisfying Assumption 5.1 from a geometric resolution is given in [37] Prop. 3.5. It follows from the main technical result Main Lemma 2.3 of that paper that we discuss here in a simplified form:

Proposition 5.5 *Set $A := k[y_1, \ldots, y_r]$ and $A[x] := A[x_1, \ldots, x_m]$. Set $n := r + m$ and let $V \subset \mathbb{A}^n$ be an equidimensional variety of dimension r defined by a reduced regular sequence $f_1, \ldots, f_m \subset A[x]$. Assume moreover that V satisfies Assumption 5.1. Suppose that a geometric resolution (q, w) of $Z = V \cap \{y_1 = 0, \ldots, y_r = 0\}$ associated to a linear form ℓ is given, and that $f_1, \ldots, f_m$ are polynomials of degrees bounded by d encoded by slp's of length L.*

Then there is a deterministic algorithm which computes (a slp for) the Chow form $\mathcal{C}h_V$ within complexity $(ndD)^{\mathcal{O}(1)} L$.

Idea of the proof.–

The computation of the Chow form relies on a way of writing it as a quotient of products of Chow forms of zero-dimensional varieties with respect to different base fields [37] Prop. 2.5:

$$\mathcal{C}h_V(U_0, \ldots, U_r) = \frac{\prod_{i=0}^r \mathcal{C}h_{Z_i}(U_i)}{\prod_{i=1}^r \mathcal{C}h_{Z_i}(e_i)}, \tag{5.4}$$

where $Z_0, \ldots, Z_r$ denote the zero-dimensional varieties of degree D defined as

$$Z_0 := Z = V(y_1, \ldots, y_r, f_1, \ldots, f_m) \subset \mathbb{A}^n(\overline{k})$$
$$Z_1 := V(L(U_0, (y, x)), y_2, \ldots, y_r, f_1, \ldots, f_m) \subset \mathbb{A}^n(\overline{k(U_0)})$$
$$\vdots$$
$$Z_r := V(L(U_0, (y, x)), \ldots, L(U_{r-1}, (y, x)), f_1, \ldots, f_m)$$
$$\subset \mathbb{A}^n(\overline{k(U_0, \ldots, U_{r-1})})$$

and $\mathcal{C}h_{Z_i}(e_i)$ consists on specializing the group of variables U_i on the $(i+1)$-vector of the canonical basis of k^{n+1} (which corresponds to the hyperplane y_i).

Now fix $0 \leq i \leq r$. Remember that as Z_i is zero-dimensional, $\mathcal{C}h_{Z_i} = \prod_{\xi_U \in Z_i} L(U_i, \xi_U)$.

It can be shown that for each $(0, \xi_0) \in Z$, Proposition 5.4 centered at $(e_1, \ldots, e_i)$ holds and gives back from ξ_0 (an approximation of) the unique $\xi_U \in \overline{k}[[U_0 - e_1, \ldots, U_{i-1} - e_i]]^m$ such that

$$\xi_U \in Z_i \quad \text{and} \quad \xi_U(e_1, \ldots, e_i) = \xi_0.$$

Then we proceed like in §5.5.2 (Idea of the algorithm) to recover Ch_{Z_i} which is the determinant of the diagonal matrix $\text{Diag}_{\xi_U \in Z_i}(L(U_i, \xi_U))$ from the companion matrix C_q of q which is similar to $\text{Diag}_{\xi_0 \in Z}(\ell(\xi_0))$.

A final comment concerning the algorithm: in order to avoid divisions in the computation of the polynomial Ch_V, we need to invert the denominator in the right hand side of Identity 5.4 and replace it by a formal power series. However as it is not directly invertible we need to compute its order and its graded component of lowest degree. This information also decides up to which order the approximations of the numerator and the denominator have to be computed. This is done in [37] Lem. 2.10.

Similar lifting ideas seemed to lead to a simpler algorithm to compute the Chow form of the variety: the Chow form can be written as the numerator of the independent term of a certain characteristic polynomial, which can be approximated from a good fiber using Newton method. However it is important to observe that up to now an algorithm that approximates the power series corresponding to a quotient does not yield an approximation of the numerator. That is the reason why the product formula above is so useful.

5.6 Arbitrary varieties

In this section $V \subset \mathbb{A}^n$ is an arbitrary variety. Thus V can be decomposed in the following manner:

$$V = V_0 \cup \cdots \cup V_n$$

where for $0 \leq r \leq n$, V_r is either empty or an equidimensional variety of dimension r. The degree D of V is defined, following [31], as the sum of the degrees of its irreducible, or equivalently equidimensional, components.

We suppose that V is described as the zero set of $f_1, \ldots, f_s \in k[x_1, \ldots, x_n]$ of degrees bounded by d and described by slp's of size bounded by L. In this section we deal with the question of producing an algorithm which determines (slp's for) the equidimensional components of V. These can be described by means of equations, or of geometric resolutions, or by their Chow forms.

Set descriptions of the equidimensional components can be found for instance in [14, 22] where the algorithms are for dense input representation and deterministic, with complexity of order $(sd^{n^2})^{\mathcal{O}(1)}$, or in [38, 39] for a probabilistic algorithm for slp input representation of complexity $(sd^n)^{\mathcal{O}(1)}L$. Geometric resolution algorithms and similar ones are given in [17] for the classic point of view and in [50, 51] for evaluation methods.

The evaluation methods algorithms have all more or less the same recursive structure, as in the zero-dimensional case. First they need a preparation of the input as described in §5.2.5 or similar in order to produce a good linear algebra underlying structure. Then the algorithms adds one equation at the time, computing at each level in the same manner equations for what they want modulo some extraneous factors (a consequence of the input preparation) that need to be cleaned at some point. Here we describe roughly the algorithm of [37] which computes probabilistically (a slp for) a Chow form of each equidimensional component of V.

Idea of the algorithm

The algorithm relies on three major ingredients:

- *Ingredient 1*: the generalization of Proposition 5.5 presented in [37] Main Lemma 2.3, where instead of being given a reduced regular sequence $f_1, \ldots, f_m$ we assume the weaker condition that $f_1, \ldots, f_m \in I(V)$ and that for every $\xi_0 \in Z$, the localized ideals $I(V)_{\xi_0}$ and $(f_1, \ldots, f_m)_{\xi_0}$ coincide.
- *Ingredient 2*: a bounded probability algorithm which given a Chow form of V and a polynomial f which is not a zero-divisor modulo $I(V)$ returns a Chow form of $V \cap V(f)$ [37] Lem. 3.8.
- *Ingredient 3*: a bounded probability algorithm which given a Chow form of an equidimensional variety with some components contained in a given hypersurface returns separated Chow forms for both parts [37] Lem. 3.9.

Let $V = V_0 \cup \cdots \cup V_n$ be the variety defined by the input polynomials. If $V \neq \mathbb{A}^n$, the input preparation (§5.2.5) enables us to assume that

$$V(f_1) = V_{n-1} \cup V'_{n-1}, \quad V(f_1, f_2) = (V_{n-2} \cup V_{n-1}) \cup V'_{n-2}$$

where the varieties V'_{n-1}, V'_{n-2} are equidimensional varieties of codimension 1 and 2 respectively containing all other components of V, and that f_1 satisfies the hypothesis of Ingredient 1 for V'_{n-1}. Also $V'_{n-1} \cap V(f_2) = (V_{n-2} \cup \tilde{V}_{n-2}) \cup V'_{n-2}$ where $\tilde{V}_{n-2}$ is the remaining equidimensional part of codimension 2 included in V_{n-1}.

The input of the first step is $\mathcal{C}h_{V(f_1)} = \mathcal{C}h_{V_{n-1} \cup V'_{n-1}}$ from which we compute $\mathcal{C}h_{V_{n-1}}$ and $\mathcal{C}h_{V'_{n-1}}$ by Ingredient 3 since $V_{n-1} \subset V(f_2)$ and no component of V'_{n-1} does.

The latter should be the input of next step: from the Chow form of V'_{n-1} one can compute, by Ingredient 2, the Chow form of $V'_{n-1} \cap V(f_2) = (V_{n-2} \cup \tilde{V}_{n-2}) \cup V'_{n-2}$ and then apply again Ingredient 3 to separate the Chow form of $V_{n-2} \cup \tilde{V}_{n-2}$ from that of V'_{n-2} (the Chow form of $V_{n-2} \cup \tilde{V}_{n-2}$ will be broken up in its two parts at the end by another application of Ingredient 3). However the complexity considerations introduced after Theorem 5.5 prevent that since the complexity would explode due to the recursion.

What we do is to compress the information of V'_{n-1}: from its Chow form we obtain probabilistically a geometric resolution of the zero-dimensional variety $Z_{n-1} = V'_{n-1} \cap V(x_1, \ldots, x_{n-1})$ associated to a certain linear form, and these arrays of coefficients will be the input of next step together with f_1.

The second step begins computing *again* $\mathcal{C}h_{V'_{n-1}}$ from the geometric resolution of Z_{n-1} and f_1 by application of Ingredient 1. Then it follows as explained in the previous paragraph computing the Chow forms of $V_{n-2} \cup \tilde{V}_{n-2}$ and of V'_{n-2}, and again keep aside the former and compress the information of the latter replacing it by a geometric resolution of $Z_{n-2} = V'_{n-2} \cap V(x_1, \ldots, x_{n-2})$.

All steps follow now the same pattern. At the end of this part of the algorithm one obtains a list of Chow forms of $V_{n-1}, V_{n-2} \cup \tilde{V}_{n-2}, \ldots, V_0 \cup \tilde{V}_0$ where $\tilde{V}_r$ is either empty or an equidimensional variety of dimension r included in $V_{r+1} \cup \cdots \cup V_{n-1}$ while no irreducible component of V_r is. The algorithm concludes extracting from these Chow forms the Chow forms of $V_{n-2}, \ldots, V_0$ by application of Ingredient 3.

The final result is in a simplified form:

Theorem 5.6 ([37] Th. 1) *Let $f_1, \ldots, f_s \in k[x_1, \ldots, x_n]$ be polynomials of degree bounded by d encoded by straight-line programs of length bounded by L. Set $V := V(f_1, \ldots, f_s) \subset \mathbb{A}^n$ and let $V = V_0 \cup \cdots \cup V_n$ be its minimal equidimensional decomposition. Set δ for a geometric degree of the input polynomial system.*

Then there is a bounded probability algorithm which computes (slp's for) Chow forms of $V_0, \ldots, V_n$ *within (expected) complexity* $s(n\,d\,\delta)^{\mathcal{O}(1)}L$. *Its worst case complexity is* $s(n\,d^n)^{\mathcal{O}(1)}L$.

An analogous result for the description of the equidimensional components of V by means of geometric resolutions is obtained in [50].

5.7 Applications

5.7.1 The computation of the sparse resultant

We take this application concerning the computation of a class of sparse resultants from [37].

The classical resultant $\mathrm{Res}_{n,d}$ of a system of $n+1$ generic homogeneous polynomials $f_0, \ldots, f_n$ of degree d in $n+1$ variables is a polynomial in the indeterminate coefficients $U_i = (U_{i\alpha}, \alpha), 0 \le i \le n$, of the polynomials f_i, that characterizes for which coefficients the system has a non-trivial solution. This polynomial is homogeneous of degree d^n in each set of indeterminate coefficients (U_i). Clearly the number of variables and the degree bound prevent to write (the dense encoding of) this polynomial, unless very specific cases like the resultant of two homogeneous polynomials in two variables. However a direct application of the computation of the Chow form, more precisely of Proposition 5.5 above, shows that a straight-line program for $\mathrm{Res}_{n,d}$ can be deterministically computed within complexity $(nd^n)^{\mathcal{O}(1)}$. This can be extended to compute some classes of sparse resultants.

The sparse resultant is the basic object in sparse elimination theory and has extensively been used as a tool for the resolution of polynomial equation systems (see for instance [65], [57], [18]). Several effective procedures were proposed to compute it (see e.g. [65], [10], [11]). Recently, C. D'Andrea has obtained an explicit determinantal formula which extends Macaulay's formula to the sparse case ([16]).

From the algorithmic point of view, the main assumption of sparse elimination theory is that computations should be substantially faster when the input polynomials are sparse (in the sense that their Newton polytopes are restricted). Basically, the parameters which control the sparsity are the number of variables n and the normalized volume $\mathrm{Vol}(\mathcal{A})$ of the convex hull of the set $\mathcal{A}$ of exponents (that is $n!$ times its volume with respect to the Euclidean volume form of $\mathbb{R}^n$). None of the previous algorithms computing sparse resultants is completely satisfactory, as their predicted complexity is exponential in all or some of these parameters (see [11] Cor. 12.8).

The precise definition of the (unmixed) sparse resultant is as follows:

Let $\mathcal{A} = \{\alpha_0, \ldots, \alpha_N\} \subset \mathbb{Z}^n$ be a finite set of integer vectors. We assume here that $\mathbb{Z}^n$ is generated by the differences of elements in $\mathcal{A}$. For $i = 0, \ldots, n$, let U_i be a group of variables indexed by the elements of $\mathcal{A}$, and set

$$f_i := \sum_{\alpha \in \mathcal{A}} U_{i\alpha}\, x^\alpha \ \in k[U_i][x_1^{\pm 1}, \ldots, x_n^{\pm 1}]$$

for the generic Laurent polynomial with support equal to $\mathcal{A}$. Let $W_{\mathcal{A}} \subset (\mathbb{P}^N)^{n+1} \times (\overline{k}^*)^n$ be the incidence variety of $f_0, \ldots, f_n$ in $(\overline{k}^*)^n$, that is

$$W_{\mathcal{A}} = \{(\nu_0, \ldots, \nu_n; \xi); \ F_i(\nu_i, \xi) = 0 \ \forall\, 0 \leq i \leq n\},$$

and let $\pi : (\mathbb{P}^N)^{n+1} \times (\overline{k}^*)^n \to (\mathbb{P}^N)^{n+1}$ be the canonical projection. The variety $\overline{\pi(W_{\mathcal{A}})}$ happens to be an irreducible variety of codimension 1 (see [21] Ch. 8, Prop.-Defn. 1.1), and the sparse $\mathcal{A}$-resultant $\operatorname{Res}_{\mathcal{A}}$ is defined as the unique — up to a sign — irreducible polynomial in $\mathbb{Z}[U_0, \ldots, U_n]$ which defines it. It is a multihomogeneous polynomial of degree $\operatorname{Vol}(\mathcal{A})$ in each group of variables U_i.

As this resultant coincides with the Chow form of the toric variety associated to the input set $\mathcal{A}$, the result one can obtain is the following:

Proposition 5.6 ([37] Cor. 4.2) *Let $\mathcal{A} \subset (\mathbb{N}_0)^n$ be a finite set which contains $\{0, e_1, \ldots, e_n\}$. Then there is a bounded probability algorithm which computes (a slp for) a scalar multiple of the $\mathcal{A}$-resultant $\operatorname{Res}_{\mathcal{A}}$ within (expected) complexity $(n\operatorname{Vol}(\mathcal{A}))^{\mathcal{O}(1)}$. Its worst case complexity is $(nd^n)^{\mathcal{O}(1)}$, where $d := \max\{|\alpha|\,;\, \alpha \in \mathcal{A}\}$.*

In fact, this expected polynomial behavior of the complexity is out of reach of the known and usual matrix formulations, as in all of them the involved matrices have an exponential size.

As an example, the $\mathcal{A}$-resultant $\operatorname{Res}_{\mathcal{A}}$ for

$$\mathcal{A} := \mathcal{A}(n,d) \\ = \{0, e_1, \ldots, e_n, e_1 + \cdots + e_n, 2e_1 + \cdots + 2e_n, \ldots, de_1 + \cdots + de_n\}$$

can be computed within expected complexity $(nd)^{\mathcal{O}(1)}$ since $\operatorname{Vol}(\mathcal{A}) = nd$.

It would be desirable to extend this result in order to compute general mixed resultants.

5.7.2 The computation of the ideal of a variety

We take this application from [4]. It is a well-known fact that unless for very particular situations there is not yet a good complexity algorithm to compute generators for the ideal of a variety from a set description of the variety. Most of the known algorithms rely on Gröbner bases computations, whose worst-case complexity is doubly exponential in the number of variables or at least in the dimension of the variety. The result here is the following:

Theorem 5.7 ([4] Th. 17) *Let $f_1, \ldots, f_s \in k[x_1, \ldots, x_n]$ be polynomials of degree bounded by d which define a smooth irreducible variety $V \subset \mathbb{A}^n$ of dimension r.*

Then there is a bounded probability algorithm which computes (slp's for) a set of $(n-r)(r+1)$ generators for $I(V)$, of degree bounded by $\deg V$ and within complexity $s(nd^n)^{\mathcal{O}(1)}$.

To give a rough idea of the algorithm we recall the notion of characteristic polynomial of an equidimensional, in our case moreover irreducible, variety $V \subset \mathbb{A}^n$ of dimension r and degree D:

Let as usual $U_0, \ldots, U_r$ be $r+1$ groups of $n+1$ variables $U_i := (U_{ij})_{0 \le j \le n}$, and $L(U_i, x) := U_{i0} + U_{i1}x_1 + \cdots + U_{in}x_n$. Also let $(t_0, \ldots, t_r)$ be a group of $r+1$ single variables. A *characteristic polynomial* $\mathcal{P}_V \in k[U_0, \ldots, U_r][t_0, \ldots, t_r]$ of V is defined as any defining equation of the Zariski closure of the image of the map

$$\varphi_V : \mathbb{A}^{(r+1)(n+1)} \times V \to \mathbb{A}^{(r+1)(n+1)} \times \mathbb{A}^{r+1},$$
$$(u_0, \ldots, u_r; \xi) \mapsto (u_0, \ldots, u_r; L(u_0, \xi), \ldots, L(u_r, \xi))$$

which is a hypersurface. This is a multihomogeneous polynomial of degree D in each group of variables $U_i \cup \{t_i\}$. Its degree in the group of variables $(t_0, \ldots, t_r)$ is also bounded by D.

By a result of [13], the ideal $I(V)$ of the smooth irreducible variety V is generated by the set of polynomials (of degree D) $\mathcal{P}_V(u, L(u_1, x), \ldots, L(u_r, x))$ for $u := (u_0, \ldots, u_r) \in \mathbb{A}^{(r+1)(n+1)}$. Moreover as $I(V)$ is locally a complete intersection (generated thus by $n-r$ polynomials), one can show that $I(V)$ can globally be generated by $(n-r)(r+1)$ of these polynomials.

The algorithmic aspects of this construction rely on the fact that the characteristic polynomial can be derived from the Chow form by a simple composition of variables ([46] Lem. 2.13), and on a careful choice of the localizations in order to recover a global description with bounded probability.

Acknowledgments

I wish to thank FoCM organizers for their invitation to give this talk at FoCM'02 great conference, held at the IMA, Minneapolis, during August 2002, and for making my presence possible there. Also, I would like to say that the results presented here would not have been obtained without any of the members of TERA group (http://tera.medicis.polytechnique.fr/) and especially without Joos Heintz. Finally, I am grateful to Juan Sabia and Martín Sombra for their help and comments.

References

[1] Berenstein, C.A and Yger, A. (1991). Effective Bézout identities in $\mathbb{Q}[x_1, \ldots, x_n]$, *Acta Math.* **166**, 69–120.
[2] Berenstein, C.A. and Yger, A. (1999). Residue calculus and effective Nullstellensatz, *Amer. J. Math.* **121**, 723–796.
[3] Berkowitz, S.J. (1984). On computing the determinant in small parallel time using a small number of processors, *Inform. Process. Lett.* **18**, 147–150.
[4] Blanco, C., Jeronimo, G. and Solernó, P. (2002). Computing generators of the ideal of a smooth algebraic variety, in preparation (University of Buenos Aires).
[5] Blum, L., Cucker, F., Shub, M. and Smale, S. (1998). *Complexity and real computation* (Springer).
[6] Bost, J.-B., Gillet, H. and Soule, C. (1994). Height of projective varieties and positive Green forms, *J. Amer. Math. Soc.* **7**, 903–1027.
[7] Brownawell, W.D. (1987). Bounds for the degrees in the Nullstellensatz, *Ann. of Math.* **126**, 577–591.
[8] ürgisser, P.B, Clausen, M. and Shokrollahi, M.A. (1997). *Algebraic complexity theory* (Springer).
[9] Caniglia, L., Galligo, A. and Heintz, J. (1998). Borne simplemente exponentielle pour les degrés dans le théorème des zéros sur un corps de charactéristique quelconte, *C. R. Acad. Sci. Paris* **307**, 255–258.
[10] Canny, J.F. and Emiris, I.Z. (1993). An efficient algorithm for the sparse mixed resultant, *Lect. Notes Comput. Sci.* **263**, 89–104.
[11] Canny, J.F. and Emiris, I.Z. (2000). A subdivision-based algorithm for the sparse resultant, *J. ACM* **47**, 417–451.
[12] Castro, D., Giusti, M., Heintz, J., Matera, G. and Pardo, L.M. (2002). The hardness of polynomial equation solving, to appear in *Foundations of Computational Mathematics*.
[13] Catanese, F. (1992). Chow varieties, Hilbert schemes and moduli spaces of surfaces of general type, *J. Alg. Geometry* **1**, 561–595.

[14] Chistov, A.L. and Grigoriev, D.Y. (1983). Subexponential time solving systems of algebraic equations. *LOMI preprint E-9-83, E-10-83, Steklov Institute, Leningrad.*

[15] Dalbec, J. and Sturmfels, B. (1995). Introduction to Chow forms, *Invariant methods in discrete and computational geometry* (Curaçao, 1994), Kluwer, 37–58.

[16] D'Andrea, C. (2001). Macaulay's style formulas for sparse resultants, E-print: math.AG/0107181.

[17] Elkadi, M. and Mourrain, B. (1999). A new algorithm for the geometric decomposition of a variety, *Proc. ISSAC'1999 (ACM)*, 9–16.

[18] Elkadi, M. and Mourrain, B. (1999). Matrices in elimination theory, *J. Symb. Comput.* **28**, 3–44.

[19] Fitchas, N. and Galligo, A. (1990). Nullstellensatz effectif et conjecture de Serre (théorème de Quillen–Suslin) pour le Calcul Formel, *Math. Nachr.* **149**, 231–253.

[20] Fitchas, N., Giusti, M. and Smietanski, F. (1995). Sur la complexité du théorème des zéros, in Gudat, J., et. al., eds., *Approximation and optimization* **8** (Peter Lange Verlag), 247–329.

[21] Gelfand, I.M., Kapranov, M.M. and Zelevinsky, A.V. (1994). *Discriminants, resultants, and multidimensional determinants* (Birkhäuser).

[22] Giusti, M. and Heintz, J. (1991). Algorithmes — disons rapides — pour la décomposition d'une varieté algébrique en composantes irréductibles et équidimensionelles, *Progress in Math.* **94** (Birkhäuser), 164–194.

[23] Giusti, M. and Heintz, J. (2001). Kronecker's smart, little black-boxes, *Proceedings of Foundations of Computational Mathematics*, Oxford 1999 (FoCM'99), Iserles, A. and DeVore, R., eds., (Cambridge University Press) **284**, 69–104.

[24] Giusti, M., Heintz, J., Hägele, K., Morais, J.E., Pardo, L.M. and Montaña, J.L. (1997). Lower bounds for Diophantine approximations, *J. Pure Appl. Algebra* **117 & 118**, 277–317.

[25] Giusti, M., Heintz, J., Morais J.E., Morgenstern, J. and Pardo, L.M. (1998). Straight-line programs in geometric elimination theory, *J. Pure Appl. Algebra* **124**, 101–146.

[26] Giusti, M., Heintz, J., Morais, J.E. and Pardo, L.M. (1995). When polynomial equation systems can be solved fast?, (Proc. 11th. International Symposium Applied Algebra, Algebraic Algorithms and Error-Correcting Codes, AAECC-11, Paris 1995, Cohen, G., Giusti, M. and Mora, T., eds.) *Lecture Notes in Comput. Sci.* **948**, 205–231.

[27] Giusti, M., Heintz, J. and Sabia, J. (1993). On the efficiency of effective Nullstellensätze, *Comput. Complexity* **3**, 56–95.

[28] Giusti, M., Lecerf, G. and Salvy, B. (2001). A Gröbner free alternative for polynomial system solving, *J. Complexity* **17**, 154–211.

[29] Hägele, K. (1998). *Intrinsic height estimates for the Nullstellensatz* (Ph.D. dissertation, Univ. Cantabria).

[30] Hägele, K., Morais, J.E., Pardo, L.M. and Sombra, M. (2000). On the intrinsic complexity of the arithmetic Nullstellensatz, *Pure Appl. Algebra* **146**, 1083–183.

[31] Heintz, J. (1983). Definability and fast quantifier elimination in algebraically closed fields, *Theoret. Comput. Sci.* **24**, 239–277.

[32] Heintz, J., Krick, T., Puddu, S., Sabia, J. and Waissbein, A. (2000). Deformation techniques for efficient polynomial equation solving, *J. Complexity* **16**, 70–109.

[33] Heintz, J., Matera, G. and Waissbein,A. (2001). On the time-space complexity of geometric elimination procedures, *Applicable Algebra in Engineering, Communication and Computing* **11** (4), 239–296.

[34] Heintz, J. and Schnorr, C.P. (1982). Testing polynomials which are easy to compute, *Proc. 812h Annual ACM Symp. on Computing* (1980) 262-268. Also in Logic and Algorithmic. An International Symposium held in honour of Ernst Specker, Monographie No. **30** de l'Enseignement de Mathématiques, Genève, 237–254.

[35] Heintz, J. and Sieveking, S. (1981). Absolute primality of polynomials is decidable in random polynomial time in the number of the variables, (8th International Colloquium on Automata, Languages and Programming ICALP 81) *Springer LN Comput. Sci.* **115**, 16–28.

[36] Hermann, G. (1926). Der Frage der endlich vielen Schritte in der Theorie der Polynomideale, *Math. Ann.* **95**, 736–788.

[37] Jeronimo, G., Krick, T., Sabia, J. and Sombra, M. (2002). The computational complexity of the Chow form, preprint (University of Buenos Aires, University of La Plata and University of Paris 7).

[38] Jeronimo, G. and Sabia, J. Probabilistic equidimensional decomposition, *C. R. Acad. Sci. Paris* **331**, 485–490.

[39] Jeronimo, G. and Sabia, J. (2002). Effective equidimensional decomposition of affine varieties, *J. Pure Appl. Algebra* **169**, 229–248.

[40] Kaltofen, E. (1985). Polynomial-time reductions from multivariate to bi- and univariate integral polynomial factorization, *SIAM J. Comput.* **14** (2), 469–489.

[41] Kaltofen, E. (1988). it Greatest common divisors of polynomials given by straight-line programs, *J. ACM* **35**, No. 1, 234–264.

[42] Kaltofen, E. (1989). Factorization of polynomials given by straight-line programs, *Randomness in Computation, Advances in Computing Research* **5** (S. Micali, ed., JAI Press Inc., CT.), 375–412.

[43] Kollár, J. (1988). Sharp effective Nullstellensatz, *J. Amer. Math. Soc.* **1**, 963–975.

[44] Krick, T. and Pardo, L.M. (1994). Une approche informatique pour l'approximation diophantienne, *C. R. Acad. Sci. Paris* **318**, 407–412.

[45] Krick, T. and Pardo, L.M. (1996). A computational method for Diophantine approximation, *Progress in Math.* **143** (Birkhäuser), 193–253.

[46] Krick, T., Pardo, L.M. and Sombra, M. (2001). Sharp estimates for the arithmetic Nullstellensatz, *Duke Math. J.* **109**, No. 3, 521–598.

[47] Krick, T., Sabia, J. and Solernó, P. (1997). On intrinsic bounds in the Nullstellensatz, *AAECC J.* **8**, 125–134.

[48] Kunz, E. (1986). Kähler differentials, *Adv. Lect. in Math.* (Vieweg-Verlag).

[49] Lecerf, G. (2000). Kronecker 0.16beta-2, April 2000, http://kronecker.medicis.polytechnique.fr/.

[50] Lecerf, G. (2000). Computing an equidimensional decomposition of an algebraic variety by means of geometric resolutions, *Proc. ISSAC'2000 (ACM)*.

[51] Lecerf, G. (2001). *Une alternative aux méthodes de réécriture pour la résolution des systèmes algébriques* (PhD Thesis, Ecole Polytechnique).

[52] Lecerf, G. (2002). Quadratic Newton iteration for systems with multiplicity, *Journal of FoCM* **2** (3), 247–293.

[53] Lecerf, G. (2002). Computing the equidimensional decomposition of an algebraic closed set by means of lifting fibers, preprint (Universit de Versailles St-Quentin-en-Yvelines).

[54] Masser, D.W. and Wüstholz, G. (1983). Fields of large transcendence degree generated by values of elliptic functions, *Invent. Math.* **72**, 407–464.

[55] Philippon, P. (1990). Dénominateurs dans le théorème des zeros de Hilbert, *Acta Arith.* **58**, 1-25.

[56] Philippon, P. Sur des hauteurs alternatives, I, *Math. Ann.* **289** (1991), 255–283; II, *Ann. Inst. Fourier* **44** (1994), 1043–1065; III, *J. Math. Pures Appl.* **74**, 345–365.

[57] Rojas, M. (1997). Toric laminations, sparse generalizd characteristic polynomials, and a refinement of Hilbert's tenth problem, *Proc. FoCM'97* (Springer-Verlag), 369–381.

[58] Sabia, J. and Solernó, P. Bounds for traces in complete intersections and degrees in the Nullstellensatz, *AAECC J.* **6**, 353–376.

[59] Schwartz, J.T. (1980). Fast probabilistic algorithms for verification of polynomial identities, *J. ACM* **27**, 701–717.

[60] Shafarevich, I. (1972). *Basic algebraic geometry* (Springer-Verlag).

[61] Sombra, M. (1997). Bounds for the Hilbert function of polynomial ideals and for the degrees in the Nullstellensatz, *J. Pure Appl. Algebra* **117** & **118**, 565–599.

[62] Sombra, M. (1999). A sparse effective Nullstellensatz, *Adv. Appl. Math.* **22**, 271–295.

[63] Sombra, M. (2002). Minima successifs de variétés toriques projectives, preprint (Univ. Paris 7).

[64] Strassen, V. (1973). Vermeidung von Divisionen, *J. Reine Angew. Math.* **264**, 182–202.

[65] Sturmfels, B. (1993). Sparse elimination theory, in D. Eisenbud and L. Robbiano, eds., *Computational algebraic geometry and commutative algebra* (Cambridge Univ. Press), 377–396.

[66] Trinks, W. (1985). On improving approximate results of Buchberger's algorithm by Newton's method, *L. N. Comput. Sci.* (Springer-Verlag) **204**, 608–611.

[67] Winkler, F.(1988). A *p*-adic approach to the computation of Gröbner bases, *J. Symb. Comput.* **6**, 287–304.

[68] Zippel, R. (1979). Probabilistic algorithms for sparse polynomials (Proceedings EUROSAM'79), *Lecture Notes in Comput. Sci.* (Springer-Verlag) **72**, 216–226.

6

Numerical Solution of Large Scale Structured Polynomial or Rational Eigenvalue Problems

Thomas Apel
Fakultät für Mathematik
Technische Universität Chemnitz
D-09107 Chemnitz, Fed. Rep. Germany

Volker Mehrmann
Institut für Mathematik, MA 4-5
Technische Universität Berlin
D-10623 Berlin, Fed. Rep. Germany

David Watkins
Department of Mathematics
Washington State University
Pullman, WA, 99164-3113, USA

Abstract

This paper deals with the numerical solution of large scale polynomial or rational eigenvalue problems with Hamiltonian or symplectic symmetry in the spectrum. Applications where such problems arise are introduced briefly. It is shown how these problems may be formulated as linear generalized eigenvalue problems that have either symmetric/skew symmetric, skew Hamiltonian/Hamiltonian or symplectic pencils. The presented numerical methods are designed to preserve these structures.

6.1 Introduction

In this paper we discuss numerical methods for the solution of large scale *nonlinear eigenvalue problems*

$$P(\lambda)v = \left(\sum_{i=0}^{k} \lambda^{i-l} M_i \right) v = 0, \qquad (6.1)$$

[0] The first author was supported by Deutsche Forschungsgemeinschaft within Project: AP 72/1-1. The second author was supported by Deutsche Forschungsgemeinschaft within Project: ME 790/14-1 and through DFG Research Center FZT86, 'Mathematics for key technologies' in Berlin.

where l is some integer. Here, the coefficients M_i are real or complex matrices, and if $l = 0$ we have a polynomial eigenvalue problem, while if $l > 0$ then we have a rational eigenvalue problem. Eigenvalue problems of this form arise in a number of applications. We will present examples in §6.2. Typically one is interested in a small number of eigenvalues and associated eigenvectors. Polynomial eigenvalue problems (in particular quadratic problems) have recently received a lot of interest in the numerical analysis community, see for example the recent survey [30]. In this paper we are, in particular, interested in two special classes of such eigenvalue problems.

The first class consists of those polynomials, where the coefficient matrices are real or complex and form alternating sequences of real symmetric and real skew-symmetric matrices, or complex symmetric and skew-symmetric matrices, i.e., $M_i^T = (-1)^i M_i$ or $M_i^T = (-1)^{i+1} M_i$ for $i = 0, \ldots, k$. The matrices in this class have a nice symmetry in the spectrum.

Proposition 6.1 ([26]) *Consider the polynomial eigenvalue problem (6.1) with alternating (real or complex) coefficients, i.e., either $M_i^T = (-1)^i M_i$ or $M_i^T = (-1)^{i+1} M_i$ for $i = 0, \ldots, k$. Then $P(\lambda)v = 0$ if and only if $v^T P(-\lambda) = 0$.*

This means that v is a right eigenvector of P associated with eigenvalue λ if and only if v^T is a left eigenvector of P associated with eigenvalue $-\lambda$. This symmetry in the spectrum is called a *Hamiltonian eigensymmetry*, i.e., the spectrum consists of quadruplets $(\lambda, -\lambda, \bar{\lambda}, -\bar{\lambda})$ in the real case or pairs $(\lambda, -\lambda)$ in the complex case if the eigenvalues are not purely imaginary.

The second class of problems consists of rational eigenvalue problems with symmetrically placed coefficients of the form

$$P_1(\kappa)v = \left[M_0 + \sum_{j=1}^{k} \left(\frac{1}{\kappa^j} M_j + \kappa^j M_j^T \right) \right] v = 0 \qquad (6.2)$$

where the coefficients are again real or complex matrices with $M_0 = M_0^T$.

The following simple result is the analogue of Proposition 6.1.

Proposition 6.2 *Consider an eigenvalue problem $P_1(\kappa)v = 0$ of the form (6.2). Then $P_1(\kappa)v = 0$ if and only if $v^T P_1(\frac{1}{\kappa}) = 0$.*

Proof The result follows by setting $\lambda = \kappa^{-1}$ and transposing. □

6. Numerical Solution of Structured Problems

We have that v is a right eigenvector of P_1 associated with eigenvalue κ if and only if v^T is a left eigenvector of P_1 associated with eigenvalue $\frac{1}{\kappa}$. This symmetry in the spectrum is called a *symplectic eigensymmetry*, i.e., the spectrum consists of quadruplets $(\kappa, \kappa^{-1}, \bar{\kappa}, \bar{\kappa}^{-1})$ in the real case or pairs (κ, κ^{-1}) in the complex case if κ is not on the unit circle.

For the described classes of eigenvalue problems the following important questions are of interest and only partially solved as of today.

- How can we design efficient numerical methods to compute several eigenvalues and associated eigenvectors (possibly from the interior of the spectrum)?
- How can we make maximal use of the symmetry structure of the coefficients and the symmetry of the spectrum?

In the next section we briefly describe some applications that lead to eigenvalue problems of the described structures and discuss these structures. Unfortunately there are essentially no viable numerical methods available, that work directly on the data of polynomial or rational eigenvalue problems to compute the desired part of the spectrum. In §6.3 we therefore discuss the classical concept of *linearization*, i.e. of transforming the high degree polynomial or rational problem into a linear eigenvalue problem of higher dimension. Since our problems have a specific structure, it would be ideal if we had linearizations with analogous structure. We show how to obtain linear eigenvalue problems with the same structures and eigensymmetries. Then in §6.4 we present numerical algorithms that exploit these structures, and that guarantee that the symmetry in the spectrum is also preserved even in finite precision arithmetic.

6.2 Applications

In order to demonstrate the importance of eigenvalue problems with Hamiltonian or symplectic spectra we will present in this section several applications, where such problems arise. We begin with two Hamiltonian examples.

Example 6.1 The study of corner singularities in anisotropic elastic materials [1, 2, 16, 19, 25, 27] leads to real quadratic eigenvalue problems of the form

$$\lambda^2 M(p)v + \lambda G(p)v + K(p)v = 0,$$
$$M = M^T, \; G = -G^T, \; K = K^T \quad (6.3)$$

The coefficient matrices are large and sparse, having been produced by a finite element discretization. M is a positive definite mass matrix, and $-K$ is a stiffness matrix. Typically the coefficients depend on a set of material or geometry parameters p which are varied.

Example 6.2 ([26]) The optimality conditions for the optimal control problem to minimize the cost functional

$$\int_{t_0}^{t_1} \sum_{i=0}^{k} (q^{(i)})^T Q_i q^{(i)} + u^T R u \, dt$$

with $Q_i = Q_i^T$, $i = 0, \ldots, k$, positive semidefinite and $R = R^T$ positive definite, subject to the control system

$$\sum_{i=0}^{k} M_i q^{(i)} = Bu(t), \tag{6.4}$$

with control input $u(t)$ and initial conditions

$$q^{(i)}(t_0) = q_{i,0}, \quad i = 0, 1, \ldots, k-1, \tag{6.5}$$

lead to the polynomial two-point boundary value problem

$$\sum_{j=1}^{k-1} \begin{bmatrix} (-1)^{j-1} Q_j & M_{2j}^T \\ M_{2j} & 0 \end{bmatrix} \begin{bmatrix} q^{(2j)} \\ \mu^{(2j)} \end{bmatrix} + \\ \sum_{j=1}^{k-1} \begin{bmatrix} 0 & -M_{2j+1}^T \\ M_{2j+1} & 0 \end{bmatrix} \begin{bmatrix} q^{(2j+1)} \\ \mu^{(2j+1)} \end{bmatrix} + \\ \begin{bmatrix} -Q_0 & M_0^T \\ M_0 & -BR^{-1}B^T \end{bmatrix} \begin{bmatrix} q \\ \mu \end{bmatrix} = 0, \tag{6.6}$$

with initial conditions (6.5) and $\mu^{(i)}(t_1) = 0$ for $i = 0, \ldots k-1$. Here we have (for simpler notation) introduced the virtual coefficients $M_{k+1} = M_{k+2} = \ldots = M_{2k} = 0$. We observe that all coefficients of derivatives higher than k are singular. If the weighting matrices Q_i are chosen to be zero for all $i \geq k/2$, then we have that all coefficients of derivatives higher than k vanish and (after possibly multiplying the second block row by -1) we obtain an alternating sequence of real symmetric and skew-symmetric coefficient matrices. The solution of this polynomial boundary value problem can then be obtained via the solution of the corresponding polynomial eigenvalue problem.

As a special case we may study the optimal control of linear mechanical systems, e.g. in robotics, which are governed by a second order differential equation of the form

$$M\ddot{q} + D\dot{q} + Kq = Bu,$$

where x and u are vectors of state and control variables, respectively. The task of computing the optimal control u that minimizes the cost functional

$$\int_{t_0}^{t_1} q^T Q_0 q + \dot{q}^T Q_1 \dot{q} + u^T Ru \, dt$$

leads to the system

$$\begin{bmatrix} Q_1 & M^T \\ M & 0 \end{bmatrix} \begin{bmatrix} \ddot{q} \\ \ddot{\mu} \end{bmatrix} + \begin{bmatrix} 0 & -D^T \\ D & 0 \end{bmatrix} \begin{bmatrix} \dot{q} \\ \dot{\mu} \end{bmatrix}$$
$$+ \begin{bmatrix} -Q_0 & K^T \\ K & -BR^{-1}B^T \end{bmatrix} \begin{bmatrix} q \\ \mu \end{bmatrix} = 0,$$

which is a special case of (6.6). The substitution

$$\begin{bmatrix} q \\ \mu \end{bmatrix} = e^{\lambda t} \begin{bmatrix} v \\ w \end{bmatrix}$$

then yields the quadratic eigenvalue problem

$$\left(\lambda^2 \begin{bmatrix} Q_1 & M^T \\ M & 0 \end{bmatrix} + \lambda \begin{bmatrix} 0 & -D^T \\ D & 0 \end{bmatrix} \right.$$
$$\left. + \begin{bmatrix} -Q_0 & K^T \\ K & -BR^{-1}B^T \end{bmatrix} \right) \begin{bmatrix} v \\ w \end{bmatrix} = 0. \qquad (6.7)$$

Other applications with similar structures arise for example in the analysis of gyroscopic systems [13, 17] or the control of semidiscretized (in space) parabolic partial differential equations [14, 13].

Our third example discusses an application that leads to an eigenvalue problem with symplectic eigensymmetry.

Example 6.3 In [15] the vibration of rails under the excitation arising from fast trains is studied. Partitioning the rail into a finite number of pieces between two crossties, see Fig. 6.1, and using classical finite element discretization leads under the assumption of an infinite rail to an infinite second order system of differential equations of the form

$$M\ddot{x} + D\dot{x} + Kx = F, \qquad (6.8)$$

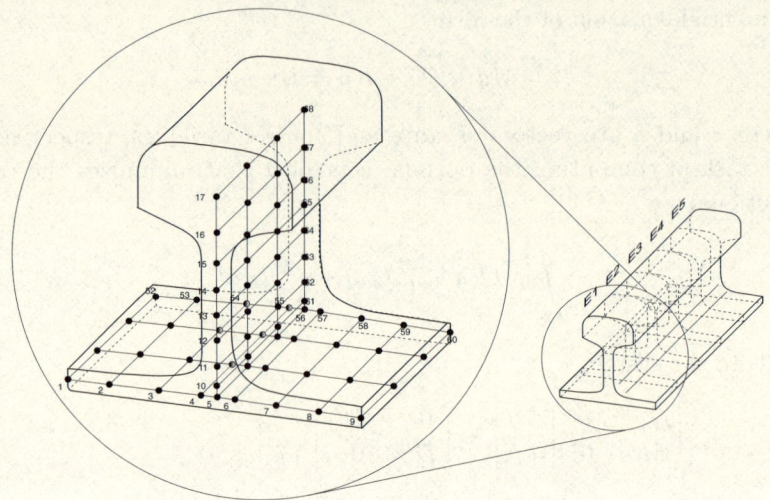

Fig. 6.1. Finite element model of rail

with infinite block tridiagonal coefficient matrices M, D, K, where

$$M = \begin{bmatrix} \ddots & \ddots & 0 & \cdots & & 0 \\ \ddots & M_{j-1,0} & M_{j,1} & 0 & & \cdots \\ 0 & M_{j,1}^T & M_{j,0} & M_{j+1,1} & & 0 \\ \vdots & \ddots & M_{j+1,1}^T & M_{j+1,0} & M_{j+2,1} \\ 0 & \cdots & 0 & & \ddots & \ddots \end{bmatrix},$$

$$x = \begin{bmatrix} \vdots \\ x_{j-1} \\ x_j \\ x_{j+1} \\ \vdots \end{bmatrix}, \quad F = \begin{bmatrix} \vdots \\ F_{j-1} \\ F_j \\ F_{j+1} \\ \vdots \end{bmatrix}.$$

The matrices D, K have the same structure with blocks $D_{j,0}, D_{j,1}$ and $K_{j,0}, K_{j,1}$, respectively. Here $M_{j,0}$ is symmetric positive definite and $D_{j,0}, K_{j,0}$ are positive semidefinite. There are several ways to approach the solution of this problem, which presents a mixture between a differential equation (time derivatives of x) and a difference equation (space differences in j).

6. Numerical Solution of Structured Problems

Since one is interested in studying the behaviour of the system under excitation, the ansatz $F_j = \hat{F}_j e^{i\omega t}$, $x_j = \hat{x}_j e^{i\omega t}$, where ω is the excitation frequency, leads to a second order difference equation with variable coefficients. The discretized system for the $\hat{x}_j$ is given by

$$A_{j,j+1}^T \hat{x}_{j-1} + A_{j,j} \hat{x}_j + A_{j,j+1} \hat{x}_{j+1} = \hat{F}_j \tag{6.9}$$

with the coefficient matrices

$$A_{j,j+1} = -\omega^2 M_{j,1} + i\omega D_{j,1} + K_{j,1}, \quad A_{j,j} = -\omega^2 M_{j,0} + i\omega D_{j,0} + K_{j,0}.$$

Observing that the system matrices vary periodically due to the identical form of the track between two crossties we may combine the (say $l = 5$) parts belonging to the rail in this space interval (see Fig. 6.1) into one vector

$$y_j = \begin{bmatrix} \hat{x}_j \\ \hat{x}_{j+1} \\ \vdots \\ \hat{x}_{j+l} \end{bmatrix}$$

and thus obtain a constant coefficient second order difference equation

$$A_1^T y_{j-1} + A_0 y_j + A_1 y_{j+1} = G_j \tag{6.10}$$

with the coefficient matrices

$$A_0 = \begin{bmatrix} A_{j,j} & A_{j,j+1} & & & & \\ A_{j+1,j+2}^T & A_{j+1,j+1} & A_{j+1,j+2} & & & \\ & \ddots & \ddots & \ddots & & \\ & & A_{j+l-1,j+l}^T & A_{j+l-1,j+l-1} & A_{j+l-1,j+l} \\ & & & A_{j+l,j+l+1}^T & A_{j+l,j+l} \end{bmatrix},$$

$$A_1 = \begin{bmatrix} 0 & 0 & \cdots & \cdots & 0 \\ 0 & \ddots & \ddots & & \\ \vdots & \ddots & \ddots & \ddots & \\ 0 & 0 & \cdots & \cdots & 0 \\ A_{j+l,j+l+1} & 0 & \cdots & \cdots & 0 \end{bmatrix}.$$

For this system we make the ansatz $y_{j+1} = \kappa y_j$, which leads to the rational eigenvalue problem

$$\left(\frac{1}{\kappa} A_1^T + A_0 + \kappa A_1 \right) y = 0.$$

Note that, since in general A_1 is singular, this is necessarily a *second order discrete time descriptor system*.

We have seen in this section that several applications lead to polynomial or rational eigenvalue problems with particular symmetry structures. In the interest of efficiency and stability, any good numerical method for solving problems of this type should preserve and exploit these structures.

6.3 Linearization

The classical approach to solving a k-th degree polynomial eigenvalue problem of dimension m is to *linearize* it, [12], i.e. to transform it to an equivalent first-degree equation $Ax - \lambda Bx = 0$ of dimension km. There are many different such linearizations with very different numerical properties, see [26, 29]. Since the problems that we have presented in §6.2 have a specific symmetry structure we will demonstrate now how we can preserve this structure in the linearization. For problems with alternating coefficients the following result has been shown in [26].

Theorem 6.1 *Consider the polynomial eigenvalue problem $P(\lambda)v = 0$ given by (6.1) with either $M_i^T = (-1)^i M_i$ or $M_i^T = (-1)^{i+1} M_i$ and with M_k nonsingular. Then the pencil $A - \lambda B \in \mathbb{C}^{mk,mk}$, where*

$$A = \begin{bmatrix} -M_0 & 0 & 0 & 0 & \cdots & 0 \\ 0 & -M_2 & -M_3 & -M_4 & \cdots & -M_k \\ 0 & M_3 & M_4 & & & 0 \\ 0 & -M_4 & & & & 0 \\ \vdots & \vdots & & & & \vdots \\ 0 & \pm M_k & 0 & 0 & \cdots & 0 \end{bmatrix} \qquad (6.11)$$

and

$$B = \begin{bmatrix} M_1 & M_2 & M_3 & \cdots & M_{k-1} & M_k \\ -M_2 & -M_3 & -M_4 & \cdots & -M_k & 0 \\ M_3 & M_4 & & & 0 & 0 \\ -M_4 & & & & 0 & 0 \\ \vdots & & & & \vdots & \vdots \\ \pm M_k & 0 & 0 & \cdots & 0 & 0 \end{bmatrix}, \qquad (6.12)$$

has the same eigenvalues as P. Here $\pm M_k$ is shorthand for $(-1)^{k-1} M_k$. If $M_i^T = (-1)^i M_i$, then A is symmetric and B is skew symmetric. If

$M_i^T = (-1)^{i+1} M_i$, then B is symmetric and A is skew symmetric. If $P(\lambda)v = 0$, then $\begin{bmatrix} v^T & \lambda v^T & \cdots & \lambda^{k-1} v^T \end{bmatrix}^T$ is an eigenvector of $A - \lambda B$.

It is obvious that the pencil $A - \lambda B$ specified by (6.11) and (6.12) is a linear eigenvalue problem, with alternating coefficients and thus by Proposition 6.1 we again have the Hamiltonian symmetry of the spectrum. If the dimension of $A - \lambda B$ is even then multiplying by J, we obtain a *skew Hamiltonian/Hamiltonian* pencil

$$\lambda H_1 - H_2 \tag{6.13}$$

with $(JH_1)^T = -JH_1$ skew Hamiltonian and $(JH_2)^T = JH_2$ Hamiltonian.

Example 6.4 If we apply Theorem 6.1 to the quadratic eigenvalue problem (6.3), we obtain the symmetric/skew-symmetric pencil

$$\begin{bmatrix} -K & 0 \\ 0 & -M \end{bmatrix} - \lambda \begin{bmatrix} G & M \\ -M & 0 \end{bmatrix}.$$

If we then multiply by $-J$, we obtain the skew Hamiltonian/Hamiltonian pencil

$$\begin{bmatrix} 0 & M \\ -K & 0 \end{bmatrix} - \lambda \begin{bmatrix} M & 0 \\ G & M \end{bmatrix}.$$

This is essentially the linearization that was used in [25].

To get a similar linearization in the symplectic case (6.2) is more difficult. First of all we have to discuss what is the analgoue to skew Hamiltonian/Hamiltonian pencils as in (6.13). Note that the discrete time analogue to Hamiltonian matrices are the *symplectic matrices* S, which satisfy $S^T J S = J$. See [22, 23, 24] for an analysis of the analogy. When it comes to pencils there are two well studied generalizations of Hamiltonian matrices, the *Hamiltonian pencils* $\lambda H_1 - H_2$ such that $H_1^T J H_2 = H_2^T J H_1$ and the skew Hamiltonian/Hamiltonian pencils (6.13).

The analogue to the Hamiltonian pencils are the *symplectic pencils* $\lambda S_1 - S_2$ such that $S_1^T J S_1 = S_2^T J S_1$ which are obtained from a Cayley transformation of Hamiltonian pencils [23, 24].

To see what is the analogue to skew Hamiltonian/Hamiltonian pencils, we perform a Cayley transformation $\lambda = \frac{\kappa-1}{\kappa+1}$ with a skew Hamiltonian/Hamiltonian pencil $\lambda H_1 - H_2$. Reordering the pencil we obtain a pencil

$\kappa(H_1 - H_2) - (H_1 + H_2) =: \kappa W_1 - W_2$ with the property that

$$W_2 = JW_1^T J^T, \tag{6.14}$$

i.e. we have a pencil of the form

$$\kappa S - JS^T J^T.$$

We call such pencils *proper symplectic pencils*.

For problems of the form (6.2) it is much more difficult to a obtain appropriate linearizations. For the second order case of Example 6.3, i.e.,

$$(\kappa^{-1}A_1^T + A_0 + \kappa A_1)y = 0. \tag{6.15}$$

we may set $\kappa z = A_1^T y$ and obtain the linear problem

$$(\kappa S_1 + S_2)\begin{bmatrix} y \\ z \end{bmatrix} := (\kappa \begin{bmatrix} A_1 & 0 \\ 0 & I \end{bmatrix} + \begin{bmatrix} A_0 & I \\ -A_1^T & 0 \end{bmatrix})\begin{bmatrix} y \\ z \end{bmatrix} = 0. \tag{6.16}$$

We immediately obtain that this linearization preserves the symplecticity of the spectrum.

Proposition 6.3 *The pencil $\kappa S_1 + S_2$ from (6.16) is a symplectic pencil, i.e. $S_1 J S_1^T = S_2 J S_2^T$. The pencil $\kappa S_1 + S_2$ is regular, i.e., its determinant does not vanish identically, if and only if the rational pencil $\kappa^{-1}A_1^T + A_0 + \kappa A_1$ is regular.*

Proof The proof of symplecticity follows by direct calculation. For the regularity observe that if the dimension of A_0 is n, then

$$\det(\kappa S_1 + S_2) = \kappa^n \det(\kappa^{-1}A_1^T + A_0 + \kappa A_1).$$ □

Note that it is not needed that the matrix A_1 is invertible to obtain a symplectic pencil, i.e. this linearization applies to Example 6.3. If, however, S_1 is invertible then clearly $S_1^{-1}S_2$ is a symplectic matrix. If A_1 is singular with rank defect $n - r$ then the pencil (6.16) has $n - r$ eigenvalues at 0 and $n - r$ eigenvalues at ∞ and hence the pairing is preserved. For more details on symplectic pencils see [11, 23].

To obtain a proper symplectic linearization we make use of the fact that a block partitioning of (6.14) gives a pencil of the form

$$\kappa \begin{bmatrix} A_{1,1} & A_{1,2} \\ A_{2,1} & A_{2,2} \end{bmatrix} - \begin{bmatrix} A_{1,1}^T & -A_{1,2}^T \\ -A_{2,1}^T & A_{1,1}^T \end{bmatrix}. \tag{6.17}$$

We can easily obtain such a linearization for (6.15) as

$$\kappa \begin{bmatrix} A_1 & A_1 \\ -A_1 & A_1^T - A_0 \end{bmatrix} - \begin{bmatrix} A_1 - A_0 & -A_1^T \\ A_1^T & A_1^T \end{bmatrix}. \qquad (6.18)$$

This is easily seen by a multiplication of (6.17) by $\begin{bmatrix} \kappa x \\ x \end{bmatrix}$ and setting this to be zero, yields the pencils

$$(\kappa^2 A_{1,1} + \kappa(A_{1,2} - A_{2,2}^T) + A_{1,2}^T)x = 0$$
$$(\kappa^2(-A_{2,1}) + \kappa(-A_{2,2} - A_{2,1}^T) + A_{1,1}^T)x = 0$$

which yields two copies of (6.15) if we choose the blocks as in (6.18).

It is much more difficult to obtain a linearization which is a symplectic pencil or a proper symplectic pencil for higher order rational problems of the form (6.2).

6.4 Numerical methods to preserve the structures

Motivated by the fact that a numerical method to solve the eigenvalue problem should preserve existing structures as much as possible, we have shown in the previous section how to transform polynomial or rational eigenvalue problems to linear generalized eigenvalue problems with the same structure.

To these problems we may then apply structure preserving methods. The development of such methods for problems with Hamiltonian or symplectic eigensymmetry has been an important research topic for many years, see [11, 23] and even though much progress has been made for problems with Hamiltonian eigensymmetry, see [4, 8, 9] for small scale methods and [1, 5, 7, 6, 25, 26, 32] for large scale problems, completely satisfactory methods, i.e. methods which are backward stable and completely structure preserving, are not yet available. This holds for both the Hamiltonian and symplectic cases.

Recently several structure-preserving Krylov subspace methods have been developed [1, 5, 6, 25, 26, 32]. Each of these requires that the Hamiltonian or symplectic pencil be reduced further to a Hamiltonian or symplectic matrix. This can be done efficiently in many cases. For example, consider the skew-symmetric/symmetric pencil in Example 6.4.

Since
$$\begin{bmatrix} G & M \\ -M & 0 \end{bmatrix}
= \begin{bmatrix} I & 0 \\ 0 & M \end{bmatrix} \begin{bmatrix} I & -\frac{1}{2}G \\ 0 & I \end{bmatrix} \begin{bmatrix} 0 & I \\ -I & 0 \end{bmatrix} \begin{bmatrix} I & 0 \\ \frac{1}{2}G & I \end{bmatrix} \begin{bmatrix} I & 0 \\ 0 & M \end{bmatrix},$$
that pencil is equivalent to the Hamiltonian matrix
$$H = J \begin{bmatrix} I & \frac{1}{2}G \\ 0 & I \end{bmatrix} \begin{bmatrix} K & 0 \\ 0 & M^{-1} \end{bmatrix} \begin{bmatrix} I & 0 \\ -\frac{1}{2}G & I \end{bmatrix},$$
as is shown in [25]. There is no need to assemble the matrix H, nor is there any need to compute M^{-1} explicitly. We just need to compute and use the Cholesky decomposition of M. Notice also that
$$H^{-1} = \begin{bmatrix} I & 0 \\ \frac{1}{2}G & 0 \end{bmatrix} \begin{bmatrix} K^{-1} & 0 \\ 0 & M \end{bmatrix} \begin{bmatrix} I & -\frac{1}{2}G \\ 0 & I \end{bmatrix} J^T,$$
so H^{-1}, which is also Hamiltonian, is no less accessible than H itself. This is important; if one wants the eigenvalues of H that are closest to the origin, one had better work with H^{-1}.

Often we prefer to shift before we invert. If we know that the eigenvalues of interest lie near τ, we might prefer to work with $(H - \tau I)^{-1}$. However, the shift destroys the Hamiltonian structure, so we need ways to effect shifts while preserving the structure. One simple remedy is to work with the matrix
$$(H - \tau I)^{-1}(H + \tau I)^{-1},$$
which is not Hamiltonian but skew Hamiltonian. If τ is neither real nor purely imaginary, we prefer to work with the skew-Hamiltonian
$$(H - \tau I)^{-1}(H - \bar{\tau} I)^{-1}(H + \tau I)^{-1}(H + \bar{\tau} I)^{-1},$$
in order to stay within the real number system.

If we wish to use a shift *and* keep the Hamiltonian structure, we can work with the Hamiltonian matrix
$$H^{-1}(H - \tau I)^{-1}(H + \tau I)^{-1}$$
or
$$H(H - \tau I)^{-1}(H + \tau I)^{-1},$$
for example.

Another possibility is to work with the Cayley transform

$$(H - \tau I)^{-1}(H + \tau I), \tag{6.19}$$

which is symplectic. Of course, the Cayley transform can also be used to transform a symplectic problem to one that is Hamiltonian. Structure-preserving Krylov subspace methods exist for each of these structures.

In practice a Krylov subspace method is never run to completion, but if it were, it would effect a similarity transformation that transforms the operator to a condensed form such as upper Hessenberg or tridiagonal. The transforming matrix is the matrix whose columns are the vectors that were generated by the process. Since in actuality a Krylov subspace method is not run to completion, it only effects a partial similarity transform.

Whether A is Hamiltonian, skew Hamiltonian, or symplectic, the structure is preserved under a symplectic similarity transformation. That is, if S is symplectic, then $S^{-1}AS$ is Hamiltonian as A is, and so on. If we write $S = [\, S_1 \; S_2 \,]$, where S_1 and S_2 are both $2n \times n$, then the symplectic property $S^T J S = J$ implies that $S_1^T J S_1 = 0$, $S_2^T J S_2 = 0$, and $S_1^T J S_2 = I$. Recall that a subspace of $\mathbb{R}^n$ is called *isotropic* if $x^T J y = 0$ for all x and y in the subspace. The condition $S_i^T J S_i = 0$ implies that the space spanned by the columns of S_i is isotropic, and so are all of its subspaces.

A Krylov subspace method will preserve Hamiltonian, skew-Hamiltonian, or symplectic structure if it generates a symplectic similarity transformation when run to completion. To achieve this it must generate vectors that span isotropic subspaces.

The skew-Hamiltonian form is easiest to preserve, since Krylov subspaces generated by skew-Hamiltonian operators are automatically isotropic [25]. Consequently the standard Arnoldi and unsymmetric Lanczos processes preserve the structure automatically in theory. In practice the isotropy is steadily eroded by roundoff errors, so it must be enforced by an additional orthogonalization step. In the context of the Arnoldi process, this means that the vector q_{j+1} generated on step j must be made orthogonal to $Jq_1, \ldots, Jq_j$ as well as $q_1, \ldots, q_j$.

All of our algorithms employ short Arnoldi or Lanczos runs with repeated implicit restarts in the spirit of Sorensen's implicitly restarted Arnoldi (IRA) process [21, 28]. The skew-Hamiltonian implicitly-restarted Arnoldi (SHIRA) algorithm [25] is a variant of IRA that includes the extra orthogonalization to enforce isotropy. SHIRA has been used in [1] for the efficient computation of corner singularities in three-dimensional anisotropic elastic structures.

A skew-Hamiltonian implicitly-restarted Lanczos (SHIRL) process can be built on the same principle. So far we have not pursued this possibility.

For the Hamiltonian case an essentially new algorithm was needed. Benner and Faßbender [5] developed a Hamiltonian Lanczos process with implicit restarts. This process has to preserve structure in both phases. In the Lanczos phase it must build isotropic subspaces, and in the restart phase it must preserve the isotropy. The latter is done by using the SR algorithm to effect the restarts, instead of the QR algorithm, which is used in IRA and SHIRA. In the algorithm of [5] there are several parameters that can be chosen freely, resulting in different variants of the algorithm. One particularly simple parameter choice leads to a simplified algorithm that uses the HR algorithm, which is much simpler than the SR algorithm, to do the restarts. This Hamiltonian implicitly restarted Lanczos (HIRL) process is outlined in [32]. HIRL has been applied successfully to the corner singularity problems of Example 6.1. It is roughly as fast and accurate as SHIRA. Since it produces both left and right eigenvectors, along with the eigenvalues, it is also able to compute residuals, as well as condition numbers for the eigenvalues.

There have been analogous developments for the symplectic case. Benner and Faßbender [6] developed a symplectic implicitly restarted Lanczos process that uses the symplectic SR algorithm for the restarts. Just as in the Hamiltonian case, it is possible to choose the parameters so that the algorithm takes a simple form that allows for restarts by the simpler HR algorithm. The resulting symplectic implicitly restarted Lanczos (SIRL) algorithm is outlined in [32]. SIRL has been tested on the singularity problems of Example 6.1, transformed to symplectic form by a Cayley transform (6.19). The algorithm was as accurate as HIRL and SHIRA but less efficient. The effectiveness of SIRL on symplectic problems of the types outlined in Example 6.3 has not yet been tested.

6.5 Numerical example

For a numerical test of some of the methods mentioned we investigate the following problem which originates from [3, 19]. Consider a specimen in form of a brick with a crack as illustrated in Fig. 6.2. The (homogeneous) material has Poisson ratio $\nu = 0.32$, the Young modulus does not influence the result and can be set arbitrarily. The stress concentration near the point O where the crack intersects the surface of the brick can

6. Numerical Solution of Structured Problems

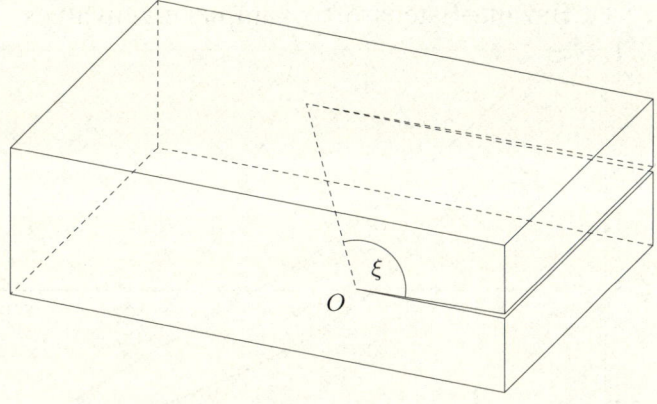

Fig. 6.2. Illustration of the crack example

be investigated within the linear elasticity framework, as long as the material is brittle, see e.g. [18, 19, 20]. The stress field is derived from the displacement field which can be represented by a regular part and several singular terms of the form $kr^\alpha u(\varphi, \theta)$ where (r, φ, θ) are spherical coordinates centered in the point of interest, k is called stress intensity factor, and α is the characteristic (singular) exponent with the associated mode u. These terms are singular when $\operatorname{Re}\alpha < 1$ and $\operatorname{Re}\alpha \neq 0$.

Mathematically, the pair (α, u) is eigenpair of a quadratic operator pencil [16, 19]. Approximation by the finite element method leads to a finite-dimensional problem, an eigenvalue problem for a quadratic matrix pencil, see [2] for discretization error estimates. After substitution $\lambda = \alpha + 0.5$ this problem has the more convenient form (6.3), see Example 6.1 in §6.2. Fig. 6.3 displays the real part of the (approximated) eigenvalues from the strip $0 < \operatorname{Re}\alpha < 1$ for the whole possible range of the angle ξ between the crack and the surface. A crack in progress has curved shape and the angle between the crack and the surface adjusts such that the corresponding eigenvalue α has real part 0.5, in our case $\xi \in \{67°, 101°, 124°\}$ where the angles are rounded to integer values. Which angle actually appears depends on the load applied. For the computation of the 2 or 3 eigenvalues of interest we computed 9 eigenvalues. Note that $\alpha = 0$ ($\lambda = 0.5$) and $\alpha = 1$ ($\lambda = 1.5$) are both triple eigenvalues corresponding to rigid body motion and without theoretical interest. We compare HIRL (with shift $\tau = 0.0$) and SHIRA with shift values $\tau = 1.0$ (useful shift in the middle of the strip of interest $0.5 < \operatorname{Re}\lambda < 1.5$) and

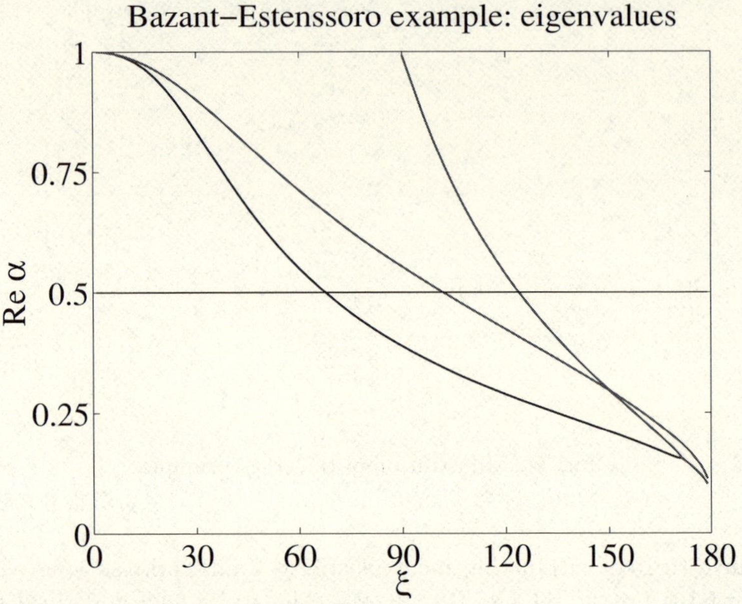

Bazant–Estenssoro example: eigenvalues

Fig. 6.3. Real part of the interesting eigenvalues against the angle ξ between the crack and the surface of the brick

$\tau = 0.0$. As a benchmark we include also a simple method that does not exploit the structure of the problem and that computes in general also eigenvalues in the left half of the complex plane if not an appropriate shift is applied; therefore we compute 18 eigenvalues in the case $\tau = 0.0$. For this test we chose $\xi = 120°$ where the three eigenvalues of interest are well separated. The total computing times with various discretization parameters are displayed in Figure 6.4 where N is the size of the matrices K, G, and M. Note that the times include the assembly of the matrices which is independent of the solution methods used.

To evaluate the algorithms we give some implementational details. The implementation of the SHIRA algorithm is based on the ARPACK package [21]. Only a slight modification is made to enforce isotropy, as discussed in §6.4. To apply the operator, we have to solve systems with the sparse matrices. In the case of SHIRA this matrix is in general ($\tau \neq 0$) non-symmetric and we use the LU-decomposition from version 2.0 of the package SuperLU [10]. In the case of HIRL we can use a Cholesky decomposition. After some comparisons we decided to use the TAUCS package for this [31]. The simple method mentioned above is

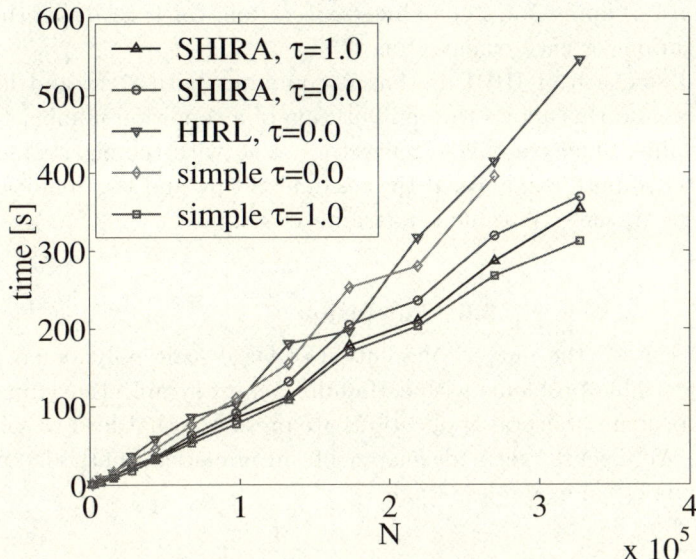

Fig. 6.4. Computing time for the crack with $\xi = 120°$ and various discretization parameters

the application of ARPACK to the matrix $(H - \tau I)^{-1}$ which has no structure. Again, we used version 2.0 of the package SuperLU for the factorization necessary. In all cases we used a stopping tolerance of 10^{-12} for the Krylov iteration. The tests were carried out on an Intel Pentium 4 CPU with 1.60 GHz, 1 GB main memory and 256 kB cache.

We conclude that SHIRA is a competitive algorithm to compute eigenvalues of quadratic skew-Hamiltonian/Hamiltonian pencils. HIRL in the current implementation is much slower which can be traced back to a much larger number of iterations in comparison with SHIRA that cannot be explained satisfactorily at this time. We are still analyzing all kinds of tolerances used in both the SHIRA and the HIRL packages. The simple method applied with the optimal shift is superior; the number of solves is nearly equal to that of SHIRA with the same shift, but some overhead, e.g. due to the additional reorthogonalization, does not occur. In extreme cases, however, we expect more stability with structure preserving methods like SHIRA.

Note that the comparison of HIRL with the other methods is not quite fair, since HIRL computes eigenvectors and both SHIRA and the simple method do not. If eigenvectors are desired these methods (in the current implementation) apply one step of inverse iteration which needs another LU factorization for each eigenvector.

A second strength of HIRL is that it computes both right and left eigenvectors, and this allows the computation of a condition number for each eigenvalue. Since we have eigenvectors to go with the eigenvalues, we can also compute residuals. If the residual is tiny and the condition number is good, the eigenvalue is guaranteed accurate.

6.6 Conclusion

We have discussed the numerical solution of large scale polynomial or rational eigenvalue problems with a Hamiltonian or symplectic symmetry in the spectrum. Several applications are presented that lead to such structures. We discuss recent developments in numerical methods and presented numerical examples.

Acknowledgments

The major work in coding the algorithms was done by Cornelia Pester, TU Chemnitz, Germany. This help is gratefully acknowledged.

References

[1] Apel, T., Mehrmann, V. and Watkins, D. (2002). Structured eigenvalue methods for the computation of corner singularities in 3D anisotropic elastic structures, *Comp. Meth. Appl. Mech. Eng.* **191**, 4459–4473.
[2] Apel, T., Sändig, A.-M. and Solov'ev, S.I. (2003). Computation of 3d vertex singularities for linear elasticity: Error estimates for a finite element method on graded meshes, *Math. Modeling Numer. Anal. (M2AN)*, page To appear.
[3] Bažant, Z.P. and Estenssoro, L.F. (1979). Surface singularity and crack propagation, *Internat. J. Solids Structures* **15**, 405–426.
[4] Benner, P., Byers, R., Mehrmann, V. and Xu, H. (2002). Numerical computation of deflating subspaces of skew Hamiltonian/Hamiltonian pencils, *SIAM J. Matrix Anal. Appl.* **24**, 165–190.
[5] Benner, P. and Faßbender, H. (1997). An implicitly restarted symplectic Lanczos method for the Hamiltonian eigenvalue problem, *Linear Algebra Appl.* **263**, 75–111.
[6] Benner, P. and Faßbender, H. (1998). An implicitly restarted symplectic Lanczos method for the symplectic eigenvalue problem (Berichte aus der Technomathematik, Report 98–01, FB3 – Mathematik und Informatik, Universität Bremen, 28334 Bremen, FRG).

[7] Benner, P. and Faßbender, H. (1998). The symplectic eigenvalue problem, the butterfly form, the *SR* algorithm, and the Lanczos method, *Linear Algebra Appl.* **275/276**, 19–47.

[8] Benner, P., Mehrmann, V. and Xu, H. (1997). A new method for computing the stable invariant subspace of a real Hamiltonian matrix, *J. Comput. Appl. Math.* **86**, 17–43.

[9] Benner, P., Mehrmann, V. and Xu, H. (1998). A numerically stable, structure preserving method for computing the eigenvalues of real Hamiltonian or symplectic pencils, *Numer. Math.* **78** (3), 329–358.

[10] Demmel, J.W., Gilbert, J.R. and Li, X.S. (1999). *SuperLU Users' Guide* (Technical Report LBNL-44289, Lawrence Berkeley National Laboratory).

[11] Faßbender, H. (2000). *Symplectic Methods for the Symplectic Eigenvalue Problem* (Kluwer Academic, New York).

[12] Gohberg, I., Lancaster, P. and Rodman, L. (1982). *Matrix Polynomials* (Academic Press, New York).

[13] Hwang, T.-M., Lin, W.-W. and Mehrmann, V. (2003) Numerical solution of quadratic eigenvalue problems for damped gyroscopic systems, *SIAM J. Sci. Comput.*, to appear.

[14] Hwang, T.-M. and Wang, W. (2002). Analyzing and visualizing a discretized semilinear elliptic problem with Neumann boundary conditions, *Numer. Methods Partial Differ. Equations* **18**, 261–279.

[15] Klimpel, T. (2003). *Verschleiß im Rad-/Schiene Kontakt infolge mittel- und hochfrequenter, dynamischer Beanspruchungen* (PhD thesis, TU Berlin, Inst. f. Luft- und Raumfahrt, in German), in preparation.

[16] Kozlov, V.A., Maz'ya, V.G. and Roßmann, J. (1997). Spectral properties of operator pencils generated by elliptic boundary value problems for the Lamé system, *Rostocker Math. Kolloq.* **51**, 5–24.

[17] Lancaster, P. (1999). Strongly stable gyroscopic systems, *Electr. J. Linear Algebra* **5**, 53–66.

[18] Lawn, B. (1993). *Fracture of Brittle Solids*, Cambridge Solid State Science Series (Cambridge University Press).

[19] Leguillon, D. (1993). Computation of 3d-singularities in elasticity, *Boundary value problems and integral equations in nonsmooth domains* (Costabel, M. et al., editor) Volume **167** of *Lect. Notes Pure Appl. Math.* (New York, Marcel Dekker), 161–170. (Proceedings of the conference, held at the CIRM, Luminy, France, May 3-7, 1993.)

[20] Leguillon, D. and Sanchez-Palencia, E. (1999). On 3d cracks intersecting a free surface in laminated composites, *Int. J. Fracture* **99**, 25–40.

[21] Lehoucq, R.B., Sorensen, D.C. and Yang, C. (1998) *ARPACK Users' Guide: Solution of Large-Scale Eigenvalue Problems with Implicitly Restarted Arnoldi Methods* (SIAM, Philadelphia).

[22] Mehl, C. (1998). *Compatible Lie and Jordan algebras and applications to structured matrices and pencils* (PhD thesis, Fakultät für Mathematik, TU Chemnitz, 09107 Chemnitz (FRG)).

[23] Mehrmann, V. (1991). The Autonomous Linear Quadratic Control Problem, Theory and Numerical Solution, *Lecture Notes in Control and Information Sciences* **163** (Springer-Verlag, Heidelberg, July 1991).

[24] Mehrmann, V. (1996). A step toward a unified treatment of continuous and discrete time control problems, *Linear Algebra Appl.* **241–243**, 749–779.

[25] Mehrmann, V. and Watkins, D. (2001). Structure-preserving methods for computing eigenpairs of large sparse skew-Hamiltoninan/Hamiltonian pencils, *SIAM J. Sci. Comput.* **22**, 1905–1925.

[26] Mehrmann, V. and Watkins, D. (2002). Polynomial eigenvalue problems with Hamiltonian structure, *Electr. Trans. Num. Anal.* **13**, 106–118.

[27] Schmitz, H., Volk, K. and Wendland, W.L. (1993). On three-dimensional singularities of elastic fields near vertices, *Numer. Methods Partial Differ. Equations* **9**, 323–337.

[28] Sorensen, D.C. (1992). Implicit application of polynomial filters in a k-step Arnoldi method, *SIAM J. Matrix Anal. Appl.* **13**, 357–385.

[29] Tisseur, F. (2000). Backward error analysis of polynomial eigenvalue problems, *Linear Algebra Appl.* **309**, 339–361.

[30] Tisseur, F. and Meerbergen, K. (2001). The quadratic eigenvalue problem, *SIAM Rev.* **43**, 234–286.

[31] Toledo, S. (2001–02). TAUCS. A library of sparse linear solvers. See http://www.math.tau.ac.il/s̃toledo/taucs/.

[32] Watkins, D.S. (2002). On Hamiltonian and symplectic Lanczos processes, *Linear Algebra Appl.*, to appear.

7

Detecting Infeasibility in Infeasible–Interior-Point Methods for Optimization

Michael J. Todd
School of Operations Research and Industrial Engineering
Cornell University
Ithaca, New York 14853, USA
Email: miketodd@cs.cornell.edu

Abstract

We study interior-point methods for optimization problems in the case of infeasibility or unboundedness. While many such methods are designed to search for optimal solutions even when they do not exist, we show that they can be viewed as implicitly searching for well-defined optimal solutions to related problems whose optimal solutions give certificates of infeasibility for the original problem or its dual. Our main development is in the context of linear programming, but we also discuss extensions to more general convex programming problems.

7.1 Introduction

The modern study of optimization began with G.B. Dantzig's formulation of the linear programming problem and his development of the simplex method in 1947. Over the more than five decades since then, the sizes of instances that could be handled grew from a few tens (in numbers of variables and of constraints) into the hundreds of thousands and even millions. During the same interval, many extensions were made, both to integer and combinatorial optimization and to nonlinear programming. Despite a variety of proposed alternatives, the simplex method remained the workhorse algorithm for linear programming, even after its non-polynomial nature in the worst case was revealed. In 1979, L.G. Khachiyan showed how the ellipsoid method of D.B. Yudin

[0] This work was supported in part by NSF through grant DMS-0209457 and ONR through grant N00014-02-1-0057.

and A.S. Nemirovskii could be applied to yield a polynomial-time algorithm for linear programming, but it was not a practical method for large-scale problems. These developments are well described in Dantzig's and Schrijver's books [4, 25] and the edited collection [18] on optimization.

In 1985, Karmarkar [9] proposed a new polynomial-time method for linear programming which did lead to practically useful algorithms, and this led to a veritable industry of developing so-called interior-point methods for linear programming problems and certain extensions. One highlight was the introduction of the concept of self-concordant barrier functions and the resulting development of polynomial-time interior-point methods for a large class of convex nonlinear programming problems by Nesterov and Nemirovskii [19]. Efficient codes for linear programming were developed, but at the same time considerable improvements to the simplex method were made, so that now both approaches are viable for very large-scale instances arising in practice: see Bixby [3]. These advances are described for example in the books of Renegar and S. Wright [24, 33] and the survey articles of M. Wright, Todd, and Forsgren et al. [32, 26, 27, 5].

Despite their very nice theoretical properties, interior-point methods do not deal very gracefully with infeasible or unbounded instances. The simplex method (a finite, combinatorial algorithm) first determines whether a linear programming instance is feasible: if not, it produces a so-called certificate of infeasibility (see §7.2.4). Then it determines whether the instance is unbounded (in which case it generates a certificate of infeasibility for the dual problem, see §7.2), and if not, produces optimal solutions for the original problem (called the primal) and its dual. By contrast, most interior-point methods (infinite iterative algorithms) assume that the instance has an optimal solution: if not, they usually give iterates that diverge to infinity, from which certificates of infeasibility can often be obtained, but without much motivation or theory. Our goal is to have a interior-point method that, in the case that optimal solutions exist, will converge to such solutions; but if not, it should produce in the limit a certificate of infeasibility for the primal or dual problem. Moreover, the algorithm should achieve this goal without knowing the status of the original problem, and in just one 'pass.'

The aim of this paper is to show that infeasible-interior-point methods, while apparently striving only for optimal solutions, can be viewed in the infeasible or unbounded case as implicitly searching for certificates of infeasibility. Indeed, under suitable conditions, the 'real' iterates produced by such an algorithm correspond to 'shadow' iterates that are generated by another interior-point method applied to a related linear programming problem whose optimal solution gives the desired certificate of infeasibility. Hence in some sense these algorithms do achieve our goal. Our main development is in the context of linear programming, but we also discuss extensions to more general convex programming problems.

§7.2 discusses linear programming problems. We define the dual problem, give optimality conditions, describe a generic primal-dual feasible-interior-point method, and discuss certificates of infeasibility. In §7.3, we describe a very attractive theoretical approach (Ye, Todd, and Mizuno [35]) to handling infeasibility in interior-point methods. The original problem and its dual are embedded in a larger self-dual problem which always has a feasible solution. Moreover, suitable optimal solutions of the larger problem can be processed to yield either optimal solutions to the original problem and its dual or a certificate of infeasibility to one of these. This approach seems to satisfy all our goals, but it does have some practical disadvantages, which we discuss.

The heart of the paper is §7.4, where we treat so-called infeasible-interior-point methods. Our main results are Theorems 7.4–7.7, which relate an interior-point iteration in the 'real' universe to one applied to a corresponding iterate in a 'shadow' universe, where the goal is to obtain a certificate of infeasibility. Thus we see that, in the case of primal or dual infeasibility, the methods can be viewed not as pursuing a chimera (optimal solutions to the primal and dual problems, which do not exist), but as implicitly following a well-defined path to optimal solutions to related problems that yield infeasibility certificates. This helps to explain the observed practical success of such methods in detecting infeasibility.

In §7.5 we discuss convergence issues. While §7.4 provides a conceptual framework for understanding the behavior of infeasible-interior-point methods in case of infeasibility, we do not have rules for choosing the parameters involved in the algorithm (in particular, step sizes) in such a way as to guarantee good progress in both the original problem and its dual *and* a suitable related problem and its dual as appropriate. We obtain results on the iterates produced by such algorithms and a convergence result (Theorem 7.8) for the method of Kojima, Megiddo, and Mizuno [10], showing that it does produce approximate certificates of infeasibility under suitable conditions.

§7.6 studies a number of interior-point methods for more general convex conic programming problems, showing (Theorem 7.9) that the results of §7.4 remain true in these settings also. We make some concluding remarks in §7.7.

7.2 Linear programming

For most of the paper, we confine ourselves to linear programming. Thus we consider the standard-form primal problem

$$(P) \text{ minimize } c^T x,$$
$$Ax = b, \quad x \geq 0,$$

of minimizing a linear function of the nonnegative variables x subject to linear equality constraints (any linear programming problem can be rewritten in this form). Closely related, and defined from the same data, is the dual problem

$$(D) \text{ maximize } b^T y,$$
$$A^T y + s = c, \quad s \geq 0.$$

Here A, an $m \times n$ matrix, $b \in \mathbb{R}^m$, and $c \in \mathbb{R}^n$ form the data; $x \in \mathbb{R}^n$ and $(y, s) \in \mathbb{R}^m \times \mathbb{R}^n$ are the variables of the problems. For simplicity, and without real loss of generality, we henceforth assume that A has full row rank.

7.2.1 Optimality conditions

If x is feasible in (P) and (y, s) in (D), then we obtain the weak duality inequality

$$c^T x - b^T y = (A^T y + s)^T x - (Ax)^T y = s^T x \geq 0, \qquad (7.1)$$

so that the objective value corresponding to a feasible primal solution is at least as large as that corresponding to a feasible dual solution. It follows that, if we have feasible solutions with equal objective values, or equivalently with $s^T x = 0$, then these solutions are optimal in their respective problems. Since $s \geq 0$ and $x \geq 0$, $s^T x = 0$ in fact implies the seemingly stronger conditions that $s_j x_j = 0$ for all $j = 1, \ldots, n$, called complementary slackness. We therefore have the following optimality conditions:

$$(OC) \quad \begin{aligned} A^T y + s &= c, & s &\geq 0, \\ Ax &= b, & x &\geq 0, \\ SXe &= 0, & & \end{aligned} \qquad (7.2)$$

where S (resp., X) denotes the diagonal matrix of order n containing the components of s (resp., x) down its diagonal, and $e \in \mathbb{R}^n$ denotes the vector of ones. These conditions are in fact necessary as well as sufficient for optimality (strong duality: see [25]).

7.2.2 The central path

The optimality conditions above consist of $m+2n$ mildly nonlinear equations in $m+2n$ variables, along with extra inequalities. Hence Newton's method seems ideal to approximate a solution, but since this necessarily has zero components, the nonnegativities cause problems. Newton's method is better suited to the following perturbed system, called the *central path equations*:

$$(CPE_\nu) \quad \begin{aligned} A^T y + s &= c, & (s > 0) \\ Ax &= b, & (x > 0) \\ SXe &= \nu e, \end{aligned} \qquad (7.3)$$

for $\nu > 0$, because if it does have a positive solution, then we can keep the iterates positive by using line searches, i.e., by employing a damped Newton method. This is the basis of primal-dual path-following methods: a few (often just one) iterations of a damped Newton method are applied to (CPE_ν) for a given $\nu > 0$, and then ν is decreased and the process continued. See, e.g., Wright [33]. We will give more details of such a method in the next subsection.

For future reference, we record the changes necessary if (P) also includes free variables. Suppose the original problem and its dual are

$$(\hat{P}) \quad \begin{aligned} \text{minimize} \quad & c^T x + d^T z, \\ & Ax + Bz = b, \quad x \geq 0, \end{aligned}$$

and

$$(\hat{D}) \quad \begin{aligned} \text{maximize} \quad & b^T y, \\ & A^T y + s = c, \quad s \geq 0, \\ & B^T y = d. \end{aligned}$$

Here B is an $m \times p$ matrix and $z \in \mathbb{R}^p$ a free primal variable. Assume that B has full column rank and $[A, B]$ full row rank, again without real loss of generality. The original problems are retrieved if B is empty.

The optimality conditions are then

$$(\widehat{OC}) \quad \begin{aligned} A^T y + s &= c, & s &\geq 0, \\ B^T y &= d, \\ Ax + Bz &= b, & x &\geq 0, \\ SXe &= 0, \end{aligned} \quad (7.4)$$

and the central path equations

$$(\widehat{CPE_\nu}) \quad \begin{aligned} A^T y + s &= c, & (s &> 0) \\ B^T y &= d, \\ Ax + Bz &= b, & (x &> 0) \\ SXe &= \nu e. \end{aligned} \quad (7.5)$$

If (7.5) has a solution, then $(\hat{P})$ and $(\hat{D})$ must have strictly feasible solutions, where the variables that are required to be nonnegative (x and s) are in fact positive. Further, the converse is true (see [33]):

Theorem 7.1 *Suppose $(\hat{P})$ and $(\hat{D})$ have strictly feasible solutions. Then, for every positive ν, there is a unique solution $(x(\nu), z(\nu), y(\nu), s(\nu))$ to (7.5). These solutions, for all $\nu > 0$, form a smooth path, and as ν approaches 0, $(x(\nu), z(\nu))$ and $(y(\nu), s(\nu))$ converge to optimal solutions to $(\hat{P})$ and $(\hat{D})$ respectively. Moreover, for every $\nu > 0$, $(x(\nu), z(\nu))$ is the unique solution to the primal barrier problem*

$$\min \quad c^T x + d^T z - \nu \sum_j \ln x_j, \quad Ax + Bz = b, \quad x > 0,$$

and $(y(\nu), s(\nu))$ the unique solution to the dual barrier problem

$$\max \quad b^T y + \nu \sum_j \ln s_j, \quad A^T y + s = c, \quad B^T y = d, \quad s > 0.$$

We call $\{(x(\nu), z(\nu)) : \nu > 0\}$ the *primal central path*, $\{(y(\nu), s(\nu)) : \nu > 0\}$ the *dual central path*, and $\{(x(\nu), z(\nu), y(\nu), s(\nu)) : \nu > 0\}$ the *primal-dual central path*.

7.2.3 A generic primal-dual feasible-interior-point method

Here we describe a simple interior-point method, leaving out the details of initialization and termination. We suppose we are solving $(\hat{P})$ and $(\hat{D})$, and that B has full column and $[A, B]$ full row rank. Let the current strictly feasible iterates be (x, z) for $(\hat{P})$ and (y, s) for $(\hat{D})$, and let μ denote $s^T x/n$. The next iterate is obtained by approximating the point on the central path corresponding to $\nu := \sigma \mu$ for some $\sigma \in [0, 1]$

7. Detecting Infeasibility

by taking a damped Newton step. Thus the search direction is found by linearizing the central path equations at the current point, so that $(\Delta x, \Delta z, \Delta y, \Delta s)$ satisfies the Newton system

$$(NS) \quad \begin{aligned} A^T \Delta y + & \Delta s = c - A^T y - s = 0, \\ B^T \Delta y & = d - B^T y = 0, \\ A\Delta x + B\Delta z & = b - Ax - Bz = 0, \\ S\Delta x & + X\Delta s = \nu e - SXe. \end{aligned} \quad (7.6)$$

Since X and S are positive definite diagonal matrices, our assumptions on A and B imply that this system has a unique solution. We then update our current iterate to

$$x_+ := x + \alpha_P \Delta x, \quad z_+ := z + \alpha_P \Delta z,$$
$$y_+ := y + \alpha_D \Delta y, \quad s_+ := s + \alpha_D \Delta s,$$

where $\alpha_P > 0$ and $\alpha_D > 0$ are chosen so that x_+ and s_+ are also positive. This concludes the iteration.

We wish to give as much flexibility to our algorithm as possible, so we will not describe rules for choosing the parameter σ and the step sizes α_P and α_D in detail. However, let us mention that, if the initial iterate is suitably close to the central path, then we can choose $\sigma := 1 - 0.1/\sqrt{n}$ and $\alpha_P = \alpha_D = 1$ and the next iterate will be strictly feasible and also suitably close to the central path. Thus these parameters can be chosen at every iteration, and this leads to a polynomial (but very slow) method; practical methods choose much smaller values for σ on most iterations. Finally, if $\alpha_P = \alpha_D$, then the duality gap $s_+^T x_+$ at the next iterate is smaller than the current one by the factor $1 - \alpha_P(1 - \sigma)$, so we would like to choose σ small and the α's large. The choice of these parameters is discussed in [33].

7.2.4 Certificates of infeasibility

In the previous subsection, we assumed that feasible, and even strictly feasible, solutions existed, and were available to the algorithm. However, it is possible that no such feasible solutions exist (often because the problem was badly formulated), and we would like to know that this is the case. Here we revert to the original problems (P) and (D), or equivalently we assume that the matrix B is null.

It is clear that, if we have $(\bar{y}, \bar{s})$ with $A^T \bar{y} + \bar{s} = 0$, $\bar{s} \geq 0$, and $b^T \bar{y} > 0$, then (P) can have no feasible solution x, for if so we would have

$$0 \geq -\bar{s}^T x = (A^T \bar{y})^T x = (Ax)^T \bar{y} = b^T \bar{y} > 0,$$

a contradiction. The well-known Farkas Lemma [25] asserts that this condition is necessary as well as sufficient:

Theorem 7.2 *The problem* (P) *is infeasible iff there exists* $(\bar{y}, \bar{s})$ *with*

$$A^T \bar{y} + \bar{s} = 0, \quad \bar{s} \geq 0, \quad \text{and} \quad b^T \bar{y} > 0. \tag{7.7}$$

We call such a $(\bar{y}, \bar{s})$ a *certificate of infeasibility for* (P).

There is a similar result for dual infeasibility:

Theorem 7.3 *The problem* (D) *is infeasible iff there exists* $\tilde{x}$ *with*

$$A\tilde{x} = 0, \quad \tilde{x} \geq 0, \quad \text{and} \quad c^T \tilde{x} < 0. \tag{7.8}$$

We call such an $\tilde{x}$ a *certificate of infeasibility for* (D). It can be shown that, if (P) is feasible, the infeasibility of (D) is equivalent to (P) being unbounded, i.e., having feasible solutions of arbitrarily low objective function value: indeed, arbitrary positive multiples of a solution $\tilde{x}$ to (7.8) can be added to any feasible solution to (P). Similarly, if (D) is feasible, the infeasibility of (P) is equivalent to (D) being unbounded, i.e., having feasible solutions of arbitrarily high objective function value.

Below we are interested in cases where the inequalities of (7.7) or (7.8) hold strictly: in this case we shall say that (P) or (D) is *strictly infeasible*. It is not hard to show, using linear programming duality, that (P) is strictly infeasible iff it is infeasible and, for every $\tilde{b}$, the set $\{x : Ax = \tilde{b}, x \geq 0\}$ is either empty or bounded, and similarly for (D). Note that, if (P) is strictly infeasible, then (D) is strictly feasible (and unbounded), because we can add any large multiple of a strictly feasible solution to (7.7) to the point $(0, c)$; similarly, if (D) is strictly infeasible, then (P) is strictly feasible (and unbounded), because we can we can add any large multiple of a strictly feasible solution to (7.8) to a point x with $Ax = b$. Finally, we remark that, if (P) is infeasible but not strictly infeasible, then an arbitrarily small perturbation to A renders (P) strictly infeasible, and similarly for (D).

7.3 The Self-dual homogeneous approach

As we mentioned in the introduction, our goal is a practical interior-point method which, when (P) and (D) are feasible, gives iterates approaching optimality for both problems; and when either is infeasible, yields a suitable certificate of infeasibility in the limit. Here we show how this can be done via a homogenization technique due to Ye, Todd, and Mizuno [35], based on work of Goldman and Tucker [6].

First consider the Goldman-Tucker system

$$
\begin{aligned}
s = &- A^T y + c\tau \geq 0, \\
Ax & - b\tau = 0, \\
\kappa = -c^T x + b^T y & \geq 0, \\
x \geq 0, \quad y \text{ free} \quad &\tau \geq 0.
\end{aligned}
\tag{7.9}
$$

This system is 'self-dual' in that the coefficient matrix is skew-symmetric, and the inequality constraints correspond to nonnegative variables while the equality constraints correspond to unrestricted variables. The system is homogeneous, but we are interested in nontrivial solutions. Note that any solution (because of the skew-symmetry) has $s^T x + \kappa\tau = 0$, and the nonnegativity then implies that $s^T x = 0$ and $\kappa\tau = 0$. If τ is positive (and hence κ zero), then scaling (x, y, s) by τ gives feasible solutions to (P) and (D) satisfying $c^T x = b^T y$, and because of weak duality, these solutions are necessarily optimal. On the other hand, if κ is positive (and hence τ zero), then either $b^T y$ is positive, which with $A^T y + s = 0$, $s \geq 0$ implies that (P) is infeasible, or $c^T x$ is negative, which with $Ax = 0$, $x \geq 0$ implies that (P) is infeasible (or both). Thus this self-dual system attacks both the optimality and the infeasibility problem together. However, it is not clear how to apply an interior-point method directly to this system.

Hence consider the linear programming problem

$$
\begin{aligned}
(HLP) \quad \min \quad & \bar{h}\theta \\
s = &- A^T y + c\tau - \bar{c}\theta \geq 0, \\
Ax & - b\tau + \bar{b}\theta = 0, \\
\kappa = -c^T x + b^T y & + \bar{g}\theta \geq 0, \\
\bar{c}^T x - \bar{b}^T y - \bar{g}\tau &\phantom{+ \bar{c}\theta} = -\bar{h}, \\
x \geq 0, \quad y \text{ free}, \quad \tau \geq 0, \quad &\theta \text{ free},
\end{aligned}
$$

where

$$\bar{b} := b\tau^0 - Ax^0, \qquad \bar{c} := c\tau^0 - A^T y^0 - s^0,$$

$$\bar{g} := c^T x^0 - b^T y^0 + \kappa^0, \qquad \bar{h} := (s^0)^T x^0 + \kappa^0 \tau^0,$$

for some initial $x^0 > 0$, y^0, $s^0 > 0$, $\tau^0 > 0$, and $\kappa^0 > 0$. Here we have added an extra artificial column to the Goldman-Tucker inequality system so that $(x^0, y^0, s^0, \tau^0, \theta^0, s^0, \kappa^0)$ is strictly feasible. To keep the skew symmetry, we also need to add an extra row. Finally, the objective function is to minimize the artificial variable θ, so as to obtain a feasible solution to (7.9).

Because of the skew symmetry, (HLP) is self-dual, i.e., equivalent to its dual, and this implies that its optimal value is attained and is zero. We can therefore apply a feasible-interior-point method to (HLP) to obtain in the limit a solution to (7.9). Further, it can be shown (see Güler and Ye [7]) that many path-following methods will converge to a strictly complementary solution, where either τ or κ is positive, and thus we can extract either optimal solutions to (P) and (D) or a certificate of infeasibility, as desired.

This technique seems to address all our concerns, since it unequivocally determines the status of the primal-dual pair of linear programming problems. However, it does have some disadvantages. First, it appears that (HLP) is of considerably higher dimension than (P), and thus that the linear system that must be solved at every iteration to obtain the search direction is of twice the dimension as that for (P). However, as long as we initialize the algorithm with corresponding solutions for (HLP) and its (equivalent) dual, we can use the self-duality to show that in fact the linear system that needs to be solved has only a few extra rows and columns compared to that for (P). Second, (HLP) links together the original primal and dual problems through the variables θ, τ, and κ, so equal step sizes must be taken in the primal and dual problems. This is definitely a drawback, since in many applications, one of the feasible regions is 'fat,' so that a step size of one can be taken without losing feasibility, while the other is 'thin' and necessitates quite small steps. There are methods allowing different step sizes [30, 34], but they are more complicated. Thirdly, only in the limit is feasibility attained, while the method of the next section allows early termination with often feasible, but not optimal, solutions.

7.4 Infeasible-interior-point methods

For the reasons just given, many codes take a simpler and more direct approach to the unavailability of initial strictly feasible solutions to (P) and (D). Lustig et al. [12, 13] proceed almost as in §7.2.3, taking a Newton step towards the (feasible) central path, but now from a point that may not be feasible for the primal or the dual. We call a triple (x, y, s) with x and s positive, but where x and/or (y, s) may not satisfy the linear equality constraints of (P) and (D), an *infeasible interior point*.

7. Detecting Infeasibility

We describe this algorithm (the infeasible-interior-point (IIP) method) precisely in the next subsection. Because its aim is to find a point on the central path, it is far from clear how this method will behave when applied to a pair of problems where either the primal or the dual is infeasible. We would like it to produce a certificate of infeasibility, but there seems little reason why it should. However, in practice, the method is amazingly successful in producing certificates of infeasibility by just scaling the iterates generated, and we wish to understand why this is. In the following subsection, we suppose that (P) is strictly infeasible, and we show that the IIP method is in fact implicitly searching for a certificate of primal infeasibility by taking damped Newton steps. Then we outline the analysis for dual strictly infeasible problems, omitting details.

7.4.1 The primal-dual infeasible-interior-point method

The algorithm described here is almost identical to the generic feasible algorithm outlined in §7.2.3. The only changes are to account for the fact that the iterates are typically infeasible interior points. For future reference, we again assume we wish to solve the more general problems $(\hat{P})$ and $(\hat{D})$, for which an infeasible interior point is a quadruple (x, z, y, s) with x and s positive.

We start at such a point (x_0, z_0, y_0, s_0). (We use subscripts for both iteration indices and components, but the latter only rarely: no confusion should arise.) At some iteration, we have a (possibly) infeasible interior point $(x, z, y, s) := (x_k, z_k, y_k, s_k)$ and, as in the feasible algorithm, we attempt to find the point on the central path corresponding to $\nu := \sigma\mu$, where $\sigma \in [0, 1]$ and $\mu := s^T x / n$, by taking a damped Newton step. The search direction is determined from

$$(NS-IIP) \quad \begin{aligned} A^T \Delta y + \Delta s &= c - A^T y - s, \\ B^T \Delta y &= d - B^T y, \\ A\Delta x + B\Delta z &= b - Ax - Bz, \\ S\Delta x + X\Delta s &= \nu e - SXe, \end{aligned} \quad (7.10)$$

whose only difference from the system (NS) is that the first three right-hand sides may be nonzero. (However, this does cause a considerable difference in the theoretical analysis, which is greatly simplified by the orthogonality of Δs and Δx in the feasible case.) Again, this system has a unique solution under our assumptions. We then update our current iterate to

$$x_+ := x + \alpha_P \Delta x, \quad z_+ := z + \alpha_P \Delta z,$$
$$y_+ := y + \alpha_D \Delta y, \quad s_+ := s + \alpha_D \Delta s,$$

where $\alpha_P > 0$ and $\alpha_D > 0$ are chosen so that x_+ and s_+ are also positive. This concludes the iteration. Note that, if it is possible to choose α_P equal to one, then (x_+, z_+) (and all subsequent primal iterates) will be feasible in $(\hat{P})$, and if α_D equals one, (y_+, s_+) (and all subsequent dual iterates) will be feasible in $(\hat{D})$.

As in the feasible case, there are many strategies for choosing the parameter σ and the step sizes α_P and α_D. Lustig et al. [12, 13] choose σ close to zero and α_P and α_D as a large multiple (say .9995) of the largest step to keep x and s positive respectively, except that steps larger than 1 are not chosen. Kojima, Megiddo, and Mizuno [10] choose a fixed $\sigma \in (0,1)$ and α_P and α_D to stay within a certain neighborhood of the central path, to keep the complementarity $s^T x$ bounded below by multiples of the primal and dual infeasibilities, and to decrease the complementarity by a suitable ratio. (More details are given in §7.5.2 below.) They are thus able to prove finite convergence, either to a point that is nearly feasible with small complementarity (and hence feasible and nearly optimal in nearby problems), or to a large enough iterate that one can deduce that there are no strictly feasible solutions to $(\hat{P})$ and $(\hat{D})$ in a large region.

Zhang [36], Mizuno [14], and Potra [23] provide extensions of Kojima et al.'s results, giving polynomial bounds to generate near-optimal solutions or guarantees that there are no optimal solutions in a large region.

These results are quite satisfactory when $(\hat{P})$ and $(\hat{D})$ are strictly feasible, but they are not as pleasant when one of these is infeasible – we would prefer to generate certificates of infeasibility, as in the method of the previous section. In the rest of this section, we show that, in the strictly infeasible case, there are 'shadow iterates' that seem to approximately indicate infeasibility. Thus in the primal infeasible case, instead of thinking of $(NS - IIP)$ as giving Newton steps towards a nonexistent primal-dual central path, we can think of it as providing a step in the shadow iterates that is a damped Newton step towards a well-defined central path for another optimization problem, which yields a primal certificate of infeasibility. This interpretation explains in some sense the practical success of infeasible-interior-point methods in detecting infeasibility.

7.4.2 The primal strictly infeasible case

Let us suppose that (P) is strictly infeasible, so that there is a solution to

$$A^T \bar{y} + \bar{s} = 0, \quad \bar{s} > 0, \quad b^T \bar{y} = 1. \tag{7.11}$$

As we showed in §7.2.4, this implies that the dual problem (D) is strictly feasible, and indeed its feasible region is unbounded. When applied to such a primal-dual pair of problems, the IIP method usually generates a sequence of iterates where (y, s) becomes feasible after a certain iteration, and $b^T y$ tends to ∞. It is easy to see that, as the iterations progress, Ax always remains a convex combination of its original value Ax_0 and its 'goal' b, but since the problem is infeasible, the weight on the first vector must remain positive. Let us therefore make the following

Assumption 7.1 *The current iterate (x, y, s) has (y, s) strictly feasible in (D) and $\beta := b^T y > 0$. In addition,*

$$Ax = \varphi Ax_0 + (1 - \varphi)b, \quad x > 0, \quad \varphi > 0.$$

If $\beta = b^T y$ is large, then $(y, s)/\beta$ will be an approximate solution to the Farkas system above. This will be part of our 'shadow iterate,' but since our IIP method is primal-dual, we also want a primal and dual for our shadow iterate. We therefore turn the Farkas system into an optimization problem, using the initial solution (x_0, y_0, s_0). Let us therefore consider

$$(\bar{D}) \max (Ax_0)^T \bar{y}$$
$$A^T \bar{y} + \bar{s} = 0,$$
$$b^T \bar{y} = 1,$$
$$\bar{s} \geq 0.$$

We call this $(\bar{D})$ since it is a homogeneous form of (D) with a normalizing constraint and a new objective function, and regard it as a dual problem of the form $(\hat{D})$. From our assumption that (P) is strictly infeasible, $(\bar{D})$ is strictly feasible. Its dual is

$$(\bar{P}) \min \quad \bar{\zeta}$$
$$A\bar{x} + b\bar{\zeta} = Ax_0,$$
$$\bar{x} \geq 0.$$

We will always use bars to indicate the variables of $(\bar{D})$ and $(\bar{P})$. Note that, from our assumption on the current iterate, $(x/\varphi, -(1-\varphi)/\varphi)$ is a strictly feasible solution to $(\bar{P})$. Hence we make the

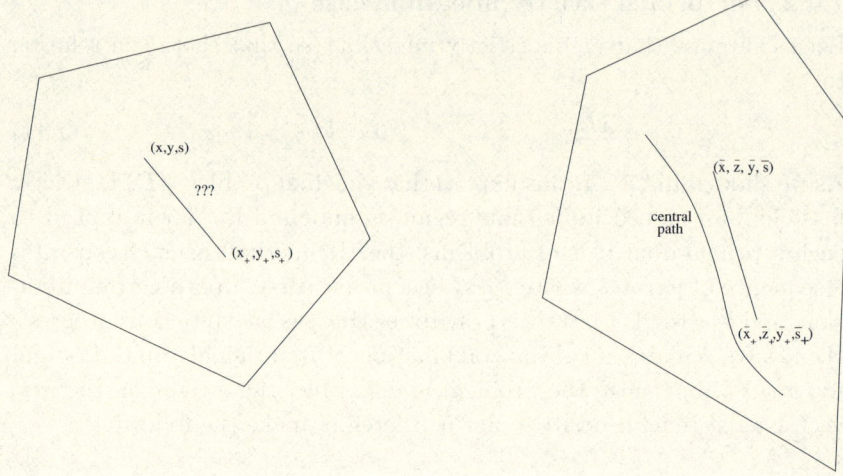

Fig. 7.1. Comparing the real and shadow iterations: a 'commutative diagram.'

Definition 7.1 *The shadow iterate corresponding to (x, y, s) is given by*

$$(\bar{x}, \bar{\zeta}) := \left(\frac{x}{\varphi}, -\frac{1-\varphi}{\varphi}\right), \quad (\bar{y}, \bar{s}) := \left(\frac{y}{\beta}, \frac{s}{\beta}\right).$$

(We note that the primal iterate x is infeasible, while the dual iterate (y, s) is feasible; these conditions are reversed in the shadow universe, where $(\bar{x}, \bar{\zeta})$ is feasible and $(\bar{y}, \bar{s})$ is typically infeasible in the first equation, while satisfying the second.)

Since φ and β are linear functions of x and (y, s) respectively, the transformations from the original iterates to the shadow iterates is a *projective* one. Projective transformations were used in Karmarkar's original interior-point algorithm [9], but have not been used much since, although they are implicit in the homogeneous approach and are used in Mizuno and Todd's analysis [15] of such methods.

We now wish to compare the results of applying one iteration of the IIP method from (x, y, s) for (P) and (D), and from $(\bar{x}, \bar{\zeta}, \bar{y}, \bar{s})$ for $(\bar{P})$ and $(\bar{D})$.

The idea is shown in the Figure 7.1. While the step from (x, y, s) to (x_+, y_+, s_+) is in some sense 'following a nonexistent central path,' the shadow iterates follow the central path for the strictly feasible pair $(\bar{P})$ and $(\bar{D})$. Indeed, the figure can be viewed as a 'commutative diagram.' Our main theorem below shows that the point $(\bar{x}_+, \bar{\zeta}_+, \bar{y}_+, \bar{s}_+)$ can be obtained either as the shadow iterate corresponding to the result of a damped Newton step for (P) and (D) from (x, y, s), or as the result of a damped Newton step for $(\bar{P})$ and $(\bar{D})$ from the shadow iterate corresponding to (x, y, s).

For a chosen value for $\sigma \in [0, 1]$, let $(\Delta x, \Delta y, \Delta s)$ be the search direction of the first of these, and let α_P and α_D be the chosen positive step sizes, with (x_+, y_+, s_+) being the next iterate. Then according to the algorithm in §7.4.1, we have

$$\begin{aligned} A^T \Delta y + \Delta s &= 0, \\ A\Delta x &= b - Ax, \\ S\Delta x + X\Delta s &= \sigma\mu e - SXe \end{aligned} \quad (7.12)$$

(note that B is empty and the dual iterate is feasible), where $\mu := s^T x/n$, and

$$x_+ := x + \alpha_P \Delta x, \quad y_+ := y + \alpha_D \Delta y, \quad s_+ := s + \alpha_D \Delta s.$$

The corresponding iteration for $(\bar{P})$ and $(\bar{D})$ also comes from §7.4.1, where now B is the single column b, but we postpone stating it until we have generated trial search directions from those above. Before doing so, we note the easily derived and well-known fact that $\Delta y = (AXS^{-1}A^T)^{-1}b - \sigma\mu(AXS^{-1}A^T)^{-1}AS^{-1}e$. Thus

$$\Delta\beta := b^T \Delta y = b^T (AXS^{-1}A^T)^{-1}b - \sigma\mu b^T (AXS^{-1}A^T)^{-1}AS^{-1}e,$$

and it follows (since infeasibility implies that b is nonzero) that $\Delta\beta$ is positive for small enough σ, depending on x and s. Henceforth, we make the

Assumption 7.2 $\Delta\beta$ *is positive.*

From Assumption 7.1, the definition of x_+, and (7.12), we find that

$$\begin{aligned} Ax_+ &= \varphi(Ax_0) + (1 - \varphi)b + \alpha_P(b - \varphi(Ax_0) - (1 - \varphi)b) \\ &= \varphi_+(Ax_0) + (1 - \varphi_+)b, \end{aligned}$$

where $\varphi_+ := (1 - \alpha_P)\varphi > 0$ (since (P) is infeasible). Also, $\beta_+ := b^T y_+ = \beta + \alpha_D \Delta\beta > 0$ from our assumptions. Hence our new shadow iterates are

$$(\bar{x}_+, \bar{\zeta}_+) := \left(\frac{x_+}{\varphi_+}, -\frac{1 - \varphi_+}{\varphi_+}\right), \quad (\bar{y}_+, \bar{s}_+) := \left(\frac{y_+}{\beta_+}, \frac{s_+}{\beta_+}\right),$$

with φ_+ and β_+ as above. We then find

$$\bar{x}_+ = \frac{x + \alpha_P \Delta x}{(1-\alpha_P)\varphi}$$

$$= \frac{x}{\varphi} + \left(\frac{\alpha_P}{1-\alpha_P} \cdot \frac{\Delta\beta}{\beta}\right)\left(\frac{\beta}{\varphi\Delta\beta}(\Delta x + x)\right)$$

$$= \bar{x} + \bar{\alpha}_P \Delta \bar{x},$$

where

$$\bar{\alpha}_P := \frac{\alpha_P}{1-\alpha_P} \cdot \frac{\Delta\beta}{\beta}, \quad \Delta\bar{x} := \frac{\beta}{\varphi\Delta\beta}(\Delta x + x), \tag{7.13}$$

and

$$\bar{\zeta}_+ = -\frac{1-(1-\alpha_P)\varphi}{(1-\alpha_P)\varphi}$$

$$= -\frac{1-\varphi}{\varphi} + \bar{\alpha}_P\left(-\frac{\beta}{\varphi\Delta\beta}\right)$$

$$= \bar{\zeta} + \bar{\alpha}_P \Delta\bar{\zeta},$$

where

$$\Delta\bar{\zeta} := -\frac{\beta}{\varphi\Delta\beta}. \tag{7.14}$$

Note that the choice of $\bar{\alpha}_P$ and hence the scale of $\Delta\bar{x}$ and $\Delta\bar{\zeta}$ is somewhat arbitrary: the particular choice made will be justified in the following theorem. Similarly the choice of $\bar{\alpha}_D$ is somewhat arbitrary below.

We also have

$$\bar{y}_+ = \frac{y + \alpha_D \Delta y}{\beta + \alpha_D \Delta \beta}$$

$$= \frac{y}{\beta} + \left(\frac{\alpha_D \Delta\beta}{\beta + \alpha_D \Delta\beta}\right)\left(\frac{\Delta y}{\Delta\beta} - \frac{y}{\beta}\right)$$

$$= \bar{y} + \bar{\alpha}_D \Delta\bar{y},$$

where

$$\bar{\alpha}_D := \frac{\alpha_D \Delta\beta}{\beta + \alpha_D \Delta\beta}, \quad \Delta\bar{y} := \frac{\Delta y}{\Delta\beta} - \bar{y}, \tag{7.15}$$

and similarly

$$\bar{s}_+ = \bar{s} + \bar{\alpha}_D \Delta\bar{s},$$

where

$$\Delta\bar{s} := \frac{\Delta s}{\Delta\beta} - \bar{s}. \tag{7.16}$$

Theorem 7.4 *The directions* $(\Delta \bar{x}, \Delta \bar{\zeta}, \Delta \bar{y}, \Delta \bar{s})$ *defined in* (7.13)–(7.16) *solve the Newton system for* $(\bar{P})$ *and* $(\bar{D})$ *given below:*

$$\begin{aligned} A^T \Delta \bar{y} + \Delta \bar{s} &= -A^T \bar{y} - \bar{s}, \\ b^T \Delta \bar{y} &= 0, \\ A \Delta \bar{x} + b \Delta \bar{\zeta} &= 0, \\ \bar{S} \Delta \bar{x} + \bar{X} \Delta \bar{s} &= \bar{\sigma} \bar{\mu} e - \bar{S} \bar{X} e, \end{aligned} \quad (7.17)$$

for the value

$$\bar{\sigma} := \frac{\beta}{\Delta \beta} \sigma. \quad (7.18)$$

Here $\bar{\mu} := \bar{s}^T \bar{x}/n$.

Proof We establish the equations of (7.17) in order. First,

$$A^T \Delta \bar{y} + \Delta \bar{s} = A^T \left(\frac{\Delta y}{\Delta \beta} - \bar{y} \right) + \left(\frac{\Delta s}{\Delta \beta} - \bar{s} \right) = -A^T \bar{y} - \bar{s},$$

using the first equation of (7.12). Next,

$$b^T \Delta \bar{y} = b^T \left(\frac{\Delta y}{\Delta \beta} - \bar{y} \right) = 1 - b^T \bar{y} = 1 - b^T y/\beta = 0$$

from the definition of $\Delta \beta$. For the third equation,

$$A \Delta \bar{x} + b \Delta \bar{\zeta} = \left(\frac{\beta}{\varphi \Delta \beta} \right) (A(\Delta x + x) - b) = 0,$$

using the second equation of (7.12). Finally, we find

$$\begin{aligned} \bar{S} \Delta \bar{x} + \bar{X} \Delta \bar{s} &= \frac{1}{\beta} \cdot \frac{\beta}{\varphi \Delta \beta} \cdot (S\Delta x + Sx) + \frac{1}{\varphi} \left(\frac{1}{\Delta \beta} X \Delta s - \frac{1}{\beta} Xs \right) \\ &= \left(\frac{\beta}{\Delta \beta} \right) \frac{1}{\beta \varphi} (S\Delta x + X\Delta s + SXe) - \frac{1}{\beta \varphi} SXe \\ &= \left(\frac{\beta}{\Delta \beta} \right) \left(\frac{1}{\beta \varphi} \sigma \mu e \right) - \frac{1}{\beta \varphi} SXe \\ &= \left(\frac{\beta}{\Delta \beta} \sigma \right) \bar{\mu} e - \bar{S} \bar{X} e, \end{aligned}$$

using the last equation of (7.12). $\square$

This theorem substantiates our main claim that, although the IIP method in the strictly infeasible case may be aiming towards a central path that doesn't exist, it is in fact implicitly trying to generate certificates of infeasibility. Indeed, the shadow iterates are being generated by damped Newton steps for the problems $(\bar{P})$ and $(\bar{D})$, for which the central path exists.

Since $(\bar{P})$ and $(\bar{D})$ are better behaved than (P) and (D), and therefore the behavior of the IIP method better understood, it is important to note that this correspondence can be reversed, to give the iteration for (P) and (D) from that for $(\bar{P})$ and $(\bar{D})$. So assume we are given $(\bar{x}, \bar{\zeta}, \bar{y}, \bar{s})$ with $A\bar{x} + b\bar{\zeta} = Ax_0$, $\bar{x} > 0$, $\bar{\zeta} \leq 0$ and $A^T\bar{y} + \bar{s} = c/\beta$, $b^T\bar{y} = 1$, $\bar{s} > 0$ for some positive β. Then we can define $\varphi := 1/(1 - \bar{\zeta}) \in (0, 1]$ so that $\bar{\zeta} = -(1-\varphi)/\varphi$, and make the

Definition 7.2 *The 'real' iterate corresponding to $(\bar{x}, \bar{\zeta}, \bar{y}, \bar{s})$ is given by*

$$x := \varphi\bar{x}, \quad (y, s) := \beta(\bar{y}, \bar{s}).$$

Thus $Ax = \varphi(Ax_0) + (-\varphi\bar{\zeta})b = \varphi(Ax_0) + (1-\varphi)b$, $x > 0$ and $A^Ty + s = c$, $s > 0$.

Suppose $(\Delta\bar{x}, \Delta\bar{\zeta}, \Delta\bar{y}, \Delta\bar{s})$ is the solution to (7.17), and also make the

Assumption 7.3 $\Delta\bar{\zeta}$ *is negative.*

This also automatically holds if $\bar{\sigma}$ is sufficiently small, and is in a sense more reasonable than Assumption 7.2 since we are now presumably close (if β is large) to a well-defined central path, and from the form of $(\bar{P})$, the assumption just amounts to monotonicity of the objective in the primal shadow problem (see Mizuno et al. [16]).

We now define our new shadow iterate $(\bar{x}_+, \bar{\zeta}_+, \bar{y}_+, \bar{s}_+)$ by taking steps in this direction, $\bar{\alpha}_P > 0$ for $(\bar{x}, \bar{\zeta})$ and $\bar{\alpha}_D > 0$ for $(\bar{y}, \bar{s})$. (We can assume that $\bar{\alpha}_D$ is less than one, since otherwise $(\bar{y}_+, \bar{s}_+)$ is a certificate of primal infeasibility for (P) and we stop.) We set $\varphi_+ := 1/(1 - \bar{\zeta}_+) = 1/(1 - \bar{\zeta} - \bar{\alpha}_P\Delta\bar{\zeta})$ (positive by Assumption 7.3) and $\beta_+ = \beta/(1 - \bar{\alpha}_D) > 0$ so that $A^T\bar{y}_+ + \bar{s}_+ = c/\beta_+$. Then we define

$$x_+ := \varphi_+\bar{x}_+ = \frac{\bar{x} + \bar{\alpha}_P\Delta\bar{x}}{1 - \bar{\zeta} - \bar{\alpha}_P\Delta\bar{\zeta}}$$

$$= x + \left(\frac{-\bar{\alpha}_P\Delta\bar{\zeta}}{1 - \bar{\zeta} - \bar{\alpha}_P\Delta\bar{\zeta}}\right)\left(\frac{\Delta\bar{x}}{-\Delta\bar{\zeta}} - x\right)$$

$$=: x + \alpha_P\Delta x$$

(i.e., α_P and Δx are defined by the expressions in parentheses in the penultimate line);

$$y_+ := \beta_+ \bar{y}_+ = \frac{\beta \bar{y} + \beta \bar{\alpha}_D \Delta \bar{y}}{1 - \bar{\alpha}_D}$$

$$= y + \left(\frac{-\bar{\alpha}_D \varphi \Delta \bar{\zeta}}{1 - \bar{\alpha}_D}\right)\left(\frac{\beta}{-\varphi \Delta \bar{\zeta}}(\Delta \bar{y} + \bar{y})\right)$$

$$=: y + \alpha_D \Delta y;$$

and similarly

$$s_+ := \beta_+ \bar{s}_+$$

$$= s + \left(\frac{-\bar{\alpha}_D \varphi \Delta \bar{\zeta}}{1 - \bar{\alpha}_D}\right)\left(\frac{\beta}{-\varphi \Delta \bar{\zeta}}(\Delta \bar{s} + \bar{s})\right)$$

$$=: s + \alpha_D \Delta s.$$

It is straightforward to check

Theorem 7.5 *The directions* $(\Delta x, \Delta y, \Delta s)$ *defined above solve the Newton system (7.12) for (P) and (D) for the value* $\sigma := \bar{\sigma}/(-\varphi \Delta \bar{\zeta})$.

We note that $b^T \Delta y = \beta/(-\varphi \Delta \bar{\zeta})$, which is positive under Assumption 7.3. This and (7.14) show that Assumptions 7.2 and 7.3 are equivalent.

The relationship between α_P and $\bar{\alpha}_P$, α_D and $\bar{\alpha}_D$, and σ and $\bar{\sigma}$ will be discussed further in the next section. For example, if we suspect that (P) is infeasible, we may want to choose α_P and α_D so that $\bar{\alpha}_P$ and $\bar{\alpha}_D$ are close to 1, so that we are taking near-Newton steps in terms of the shadow iterates.

7.4.3 The dual strictly infeasible case

Now we sketch the analysis for the dual strictly infeasible case, omitting details. We suppose there is a solution to

$$A\tilde{x} = 0, \quad \tilde{x} > 0, \quad c^T \tilde{x} = -1.$$

In this case, the IIP algorithm usually generates a sequence of iterates where x becomes feasible after a certain iteration, and $c^T x$ tends to $-\infty$. $A^T y + s$ always remains a convex combination of its original value $A^T y_0 + s_0$ and its goal c. Thus we make the following

Assumption 7.4 *The current iterate (x, y, s) has x feasible in (P) and $\gamma := -c^T x > 0$. In addition,*

$$A^T y + s = \psi(A^T y_0 + s_0) + (1 - \psi)c, \quad s > 0, \quad \psi > 0.$$

If $c^T x$ is large and negative, then x/γ will be an approximate solution to the Farkas system above. We formulate the optimization problem

$$(\tilde{P}) \quad \min \ (A^T y_0 + s_0)^T \tilde{x}$$
$$A\tilde{x} = 0,$$
$$-c^T \tilde{x} = 1,$$
$$\tilde{x} \geq 0.$$

$((\tilde{P})$ is a modified homogeneous form of (P).) This is strictly feasible. Its dual is

$$(\tilde{D}) \quad \max \quad \tilde{\kappa}$$
$$A^T \tilde{y} - c\tilde{\kappa} + \tilde{s} = A^T y_0 + s_0,$$
$$\tilde{s} \geq 0.$$

We will use tildes to indicate the variables of $(\tilde{P})$ and $(\tilde{D})$. Note that, from our assumption on the current iterate, $(y/\psi, (1-\psi)/\psi, s/\psi)$ is a strictly feasible solution to $(\tilde{D})$. Hence we make the

Definition 7.3 *The shadow iterate corresponding to (x, y, s) is given by*

$$\tilde{x} := x/\gamma, \text{ where } \gamma := -c^T x, \quad (\tilde{y}, \tilde{\kappa}, \tilde{s}) := \left(\frac{y}{\psi}, \frac{1-\psi}{\psi}, \frac{s}{\psi}\right).$$

We now wish to compare the results of applying one iteration of the IIP method from (x, y, s) for (P) and (D), and from $(\tilde{x}, \tilde{y}, \tilde{\kappa}, \tilde{s})$ for $(\tilde{P})$ and $(\tilde{D})$.

Let $(\Delta x, \Delta y, \Delta s)$ be the search direction of the first of these, and let α_P and α_D be the chosen positive step sizes, with (x_+, y_+, s_+) being the next iterate. Then according to the algorithm, we have

$$\begin{aligned} A^T \Delta y + \Delta s &= c - A^T y - s, \\ A\Delta x &= 0, \\ S\Delta x + X\Delta s &= \sigma \mu e - SXe. \end{aligned} \quad (7.19)$$

and

$$x_+ := x + \alpha_P \Delta x, \quad y_+ := y + \alpha_D \Delta y, \quad s_+ := s + \alpha_D \Delta s,$$

where $\mu := s^T x/n$. The corresponding iteration for $(\tilde{P})$ and $(\tilde{D})$ also comes from §7.4.1, where now A is augmented by the row $-c^T$, but we postpone stating it until we have generated trial search directions from those above. Before doing so, we note that

$$\Delta x = -(I - XS^{-1}A^T(AXS^{-1}A^T)^{-1}A)XS^{-1}c$$
$$+\sigma\mu(I - XS^{-1}A^T(AXS^{-1}A^T)^{-1}A)S^{-1}e,$$

and so

$$\Delta\gamma := -c^T\Delta x$$
$$= [c^T XS^{-1}c - c^T XS^{-1}A^T(AXS^{-1}A^T)^{-1}AXS^{-1}c]$$
$$-\sigma\mu(c^T S^{-1}e - c^T XS^{-1}A^T(AXS^{-1}A^T)^{-1}AS^{-1}e),$$

and it follows (since dual infeasibility implies that c is not in the range of A^T) that $\Delta\gamma$ is positive for small enough σ. Henceforth, we make the

Assumption 7.5 $\Delta\gamma$ *is positive.*

We find that

$$A^T y_+ + s_+$$
$$= \psi(A^T y_0 + s_0) + (1-\psi)c + \alpha_D(c - \psi(A^T y_0 + s_0) - (1-\psi)c)$$
$$= \psi_+(A^T y_0 + s_0) + (1-\psi_+)c,$$

where $\psi_+ := (1-\alpha_D)\psi > 0$ (since (D) is infeasible). Also, $\gamma_+ := -c^T x_+ = \gamma + \alpha_P \Delta\gamma > 0$ from our assumptions. Hence our new shadow iterates are

$$\tilde{x}_+ := \frac{x_+}{\gamma_+}, \quad (\tilde{y}_+, \tilde{\kappa}_+, \tilde{s}_+) := (\frac{y_+}{\psi_+}, \frac{1-\psi_+}{\psi_+}, \frac{s_+}{\psi_+}).$$

We then obtain

$$\tilde{x}_+ = \frac{x + \alpha_P\Delta x}{\gamma + \alpha_P\Delta\gamma}$$
$$= \frac{x}{\gamma} + \left(\frac{\alpha_P\Delta\gamma}{\gamma + \alpha_P\Delta\gamma}\right)\left(\frac{\Delta x}{\Delta\gamma} - \frac{x}{\gamma}\right)$$
$$= \tilde{x} + \tilde{\alpha}_P\Delta\tilde{x},$$

where

$$\tilde{\alpha}_P := \frac{\alpha_P\Delta\gamma}{\gamma + \alpha_P\Delta\gamma}, \quad \Delta\tilde{x} := \frac{\Delta x}{\Delta\gamma} - \tilde{x}. \tag{7.20}$$

We also have

$$\tilde{y}_+ = \frac{y + \alpha_D \Delta y}{(1 - \alpha_D)\psi}$$
$$= \frac{y}{\psi} + \left(\frac{\alpha_D}{1-\alpha_D} \cdot \frac{\Delta \gamma}{\gamma}\right)\left(\frac{\gamma}{\psi \Delta \gamma}(\Delta y + y)\right)$$
$$= \tilde{y} + \tilde{\alpha}_D \Delta \tilde{y},$$

where

$$\tilde{\alpha}_D := \frac{\alpha_D}{1-\alpha_D} \cdot \frac{\Delta \gamma}{\gamma}, \quad \Delta \tilde{y} := \frac{\gamma}{\psi \Delta \gamma}(\Delta y + y). \tag{7.21}$$

Similarly, $\tilde{s}_+ = \tilde{s} + \tilde{\alpha}_D \Delta \tilde{s}$, where

$$\Delta \tilde{s} := \frac{\gamma}{\psi \Delta \gamma}(\Delta s + s), \tag{7.22}$$

and $\tilde{\kappa}_+ = \tilde{\kappa} + \tilde{\alpha}_D \Delta \tilde{\kappa}$, where

$$\Delta \tilde{\kappa} := \frac{\gamma}{\psi \Delta \gamma}. \tag{7.23}$$

Theorem 7.6 *The directions* $(\Delta \tilde{x}, \Delta \tilde{y}, \Delta \tilde{\kappa}, \Delta \tilde{s})$ *defined in (7.20) - (7.23) above solve the Newton system for* $(\tilde{P})$ *and* $(\tilde{D})$ *given below:*

$$\begin{aligned} A^T \Delta \tilde{y} - c \Delta \tilde{\kappa} + \Delta \tilde{s} &= 0, \\ A \Delta \tilde{x} &= -A \tilde{x}, \\ -c^T \Delta \tilde{x} &= 0, \\ \tilde{S} \Delta \tilde{x} + \tilde{X} \Delta \tilde{s} &= \tilde{\sigma} \tilde{\mu} e - \tilde{X} \tilde{S} e, \end{aligned} \tag{7.24}$$

for the value

$$\tilde{\sigma} := \frac{\gamma}{\Delta \gamma} \sigma. \tag{7.25}$$

Here $\tilde{\mu} := \tilde{s}^T \tilde{x}/n$.

This argument can also be reversed. Given $(\tilde{x}, \tilde{y}, \tilde{\kappa}, \tilde{s})$, where we assume that $A\tilde{x} = b/\gamma$, $-c^T\tilde{x} = 1$, $\tilde{x} > 0$ for some positive γ and $A^T\tilde{y} - c\tilde{\kappa} + \tilde{s} = A^T y_0 + s_0$, $\tilde{s} > 0$, and $\tilde{\kappa} \geq 0$, we define $\psi := 1/(1+\tilde{\kappa}) \in (0, 1]$ so that $\tilde{\kappa} = (1 - \psi)/\psi$, and hence the 'real' iterate given by $x := \gamma \tilde{x}$, $(y, s) := \psi(\tilde{y}, \tilde{s})$. We compute the search direction from (7.24) and take steps of size $\tilde{\alpha}_P$ (assumed less than one, otherwise we have a certificate of dual infeasibility) and $\tilde{\alpha}_D$ to obtain new shadow iterates. The appropriate requirement is

Assumption 7.6 $\Delta \tilde{\kappa}$ *is positive,*

7. Detecting Infeasibility

which turns out to be equivalent to our previous assumption that $\Delta\gamma > 0$. Then the new real iterates corresponding to the new shadow iterates are obtained from the old real iterates by using the step sizes and directions given below:

$$\alpha_P := \frac{\tilde{\alpha}_P \psi \Delta \tilde{\kappa}}{1 - \tilde{\alpha}_P}, \qquad \Delta x := \frac{\gamma}{\psi \Delta \tilde{\kappa}}(\Delta \tilde{x} + \tilde{x}),$$

$$\alpha_D := \frac{\tilde{\alpha}_D \Delta \tilde{\kappa}}{1 + \tilde{\kappa} + \tilde{\alpha}_D \Delta \tilde{\kappa}}, \qquad \Delta y := \frac{\Delta \tilde{y}}{\Delta \tilde{\kappa}} - y, \qquad \Delta s := \frac{\Delta \tilde{s}}{\Delta \tilde{\kappa}} - s.$$

Again it is easy to check

Theorem 7.7 *The directions* $(\Delta x, \Delta y, \Delta s)$ *defined above solve the Newton system (7.19) for* (P) *and* (D) *for the value* $\sigma := \bar{\sigma}/(\psi \Delta \tilde{\kappa})$.

7.5 Convergence and implications

Here we give further properties of the iterates in the infeasible case, discuss the convergence of IIP methods in case of strict infeasibility, and consider the implications of our equivalence between real and shadow iterations for designing an efficient IIP method. In §7.5.1 we discuss the boundedness of the iterates in the infeasible case, while in §7.5.2 we consider the Kojima-Megiddo-Mizuno algorithm and convergence issues. Finally, §7.5.3 addresses the implications of our equivalence results for IIP methods.

7.5.1 Boundedness and unboundedness

Here we will assume that (P) is strictly infeasible, so that there is a solution to (7.11), which we repeat here:

$$A^T \bar{y} + \bar{s} = 0, \quad \bar{s} > 0, \quad b^T \bar{y} = 1.$$

(Similar results can be obtained in the dual strictly infeasible case.)

Note that any primal-dual IIP method has iterates (x_k, y_k, s_k) that satisfy

$$Ax_k = b_k := \varphi_k(Ax_0) + (1 - \varphi_k)b, \quad 0 \leq \varphi_k \leq 1, \tag{7.26}$$

and

$$A^T y_k + s_k = c_k := \psi_k(A^T y_0 + s_0) + (1 - \psi_k)c, \quad 0 \leq \psi_k \leq 1, \tag{7.27}$$

for all k.

Proposition 7.1 *In the primal strictly infeasible case, we have*

$$\varphi_k \geq (1 + \bar{s}^T x_0)^{-1}, \qquad \bar{s}^T x_k \leq \bar{s}^T x_0 \qquad (7.28)$$

for all $k \geq 0$. Hence all x_k's lie in a bounded set. Further, for any $\tilde{b}$ with $\|\tilde{b} - b\| < 1/\|\bar{y}\|$, the system $Ax = \tilde{b}, x \geq 0$ is infeasible.

Proof For the first part, premultiply (7.26) by $-\bar{y}^T$ to get

$$\begin{aligned}
\bar{s}^T x_k &= -\bar{y}^T A x_k = \varphi_k(-\bar{y}^T A x_0) + (1 - \varphi_k)(-b^T \bar{y}) \\
&= \varphi_k \bar{s}^T x_0 - 1 + \varphi_k \\
&= \varphi_k(1 + \bar{s}^T x_0) - 1.
\end{aligned}$$

Since $\bar{s}^T x_k > 0$, we obtain the lower bound on φ_k. From $\varphi_k \leq 1$, the upper bound on $\bar{s}^T x_k$ holds. For the second part, note that $\tilde{b}^T \bar{y} = b^T \bar{y} + (\tilde{b} - b)^T \bar{y} \geq 1 - \|\tilde{b} - b\| \|\bar{y}\| > 0$, so that $(\bar{y}, \bar{s})$ certifies the infeasibility of $Ax = \tilde{b}, x \geq 0$. $\square$

Proposition 7.2 *Suppose that in addition the sequence $\{(x_k, y_k, s_k)\}$ satisfies $s_k^T x_k \leq s_0^T x_0$ and $\|s_k\| \to \infty$. Then $b^T y_k \to \infty$.*

Proof Indeed, we have

$$\begin{aligned}
s_0^T x_0 \geq s_k^T x_k &= (c_k - A^T y_k)^T x_k \\
&= c_k^T x_k - y_k^T[\varphi_k(A x_0) + (1 - \varphi_k)b] \\
&= c_k^T x_k - \varphi_k(A^T y_k)^T x_0 - (1 - \varphi_k)b^T y_k \\
&= c_k^T x_k - \varphi_k(c_k - s_k)^T x_0 - (1 - \varphi_k)b^T y_k \\
&= [c_k^T x_k - \varphi_k c_k^T x_0] + \varphi_k s_k^T x_0 - (1 - \varphi_k)b^T y_k.
\end{aligned} \qquad (7.29)$$

Now, by Proposition 7.1, the quantity in brackets remains bounded, while $\varphi_k \geq (1 + \bar{s}^T x_0)^{-1} > 0$ and $s_k^T x_0 \to \infty$. Thus we must have $b^T y_k \to \infty$. $\square$

7.5.2 The Kojima-Megiddo-Mizuno algorithm and convergence

Kojima, Megiddo, and Mizuno [10] (henceforth KMM) devised a particular IIP method that correctly detected infeasibility, but without generating a certificate of infeasibility in the usual sense. Here we show that their algorithm does indeed generate certificates of infeasibility in the limit (in the strictly infeasible case). We also see how their method relates to the assumptions and shadow iterates we studied in §7.4.2.

KMM's algorithm uses special rules for choosing σ, α_P, and α_D at each iteration, and employs a special neighborhood: for (P) and (D), this is defined to be

$$\mathcal{N} := \mathcal{N}(\gamma_0, \gamma_P, \gamma_D, \varepsilon_P, \varepsilon_D) := \mathcal{N}_0 \cap \mathcal{N}_P \cap \mathcal{N}_D, \quad \text{where}$$
$$\mathcal{N}_0 := \{(x, y, s) \in \mathbb{R}^n_{++} \times \mathbb{R}^m \times \mathbb{R}^n_{++} : s_j x_j \geq \gamma_0 s^T x/n, \text{ for all } j\},$$
$$\mathcal{N}_P := \{(x, y, s) \in \mathbb{R}^n_{++} \times \mathbb{R}^m \times \mathbb{R}^n_{++} :$$
$$\|Ax - b\| \leq \max(\varepsilon_P, s^T x/\gamma_P)\},$$
$$\mathcal{N}_D := \{(x, y, s) \in \mathbb{R}^n_{++} \times \mathbb{R}^m \times \mathbb{R}^n_{++} :$$
$$\|A^T y + s - c\| \leq \max(\varepsilon_D, s^T x/\gamma_D)\}. \qquad (7.30)$$

Here ε_P and ε_D are small positive constants, and $\gamma_0 < 1$, γ_P, and γ_D are positive constants chosen so that $(x_0, y_0, s_0) \in \mathcal{N}$. KMM maintain all iterates in $\mathcal{N}$.

They choose parameters $0 < \sigma_1 < \sigma_2 < \sigma_3 < 1$. At every iteration, σ is chosen to be σ_1 to generate search directions $(\Delta x, \Delta y, \Delta s)$ from the current iterate $(x, y, s) \in \mathcal{N}$. (In fact, it suffices for their arguments to choose σ from the interval $[\sigma'_1, \sigma''_1]$, possibly with different choices at each iteration, where $0 < \sigma'_1 < \sigma''_1 < \sigma_2 < \sigma_3 < 1$.) Next, a step size $\bar{\alpha}$ is chosen as the largest $\tilde{\alpha} \leq 1$ so that

$$(x + \alpha \Delta x, y + \alpha \Delta y, s + \alpha \Delta s) \in \mathcal{N} \qquad \text{and}$$
$$(s + \alpha \Delta s)^T (x + \alpha \Delta x) \leq [1 - \alpha(1 - \sigma_2)] s^T x$$

for all $\alpha \in [0, \tilde{\alpha}]$. Finally, $\alpha_P \leq 1$ and $\alpha_D \leq 1$ are chosen so that

$$\begin{aligned}(x + \alpha_P \Delta x, y + \alpha_D \Delta y, s + \alpha_D \Delta s) \in \mathcal{N} \qquad \text{and} \\ (s + \alpha_D \Delta s)^T (x + \alpha_P \Delta x) \leq [1 - \bar{\alpha}(1 - \sigma_3)] s^T x\end{aligned} \qquad (7.31)$$

Note that a possible choice is $\alpha_P = \alpha_D = \bar{\alpha}$. However, the relaxation provided by choosing $\sigma_3 > \sigma_2$ allows other options; in particular, it might be possible to choose one of α_P and α_D as 1 (thus attaining primal or dual feasibility) while the other is necessarily small (because the dual or primal problem is infeasible).

The algorithm is terminated whenever an iterate (x, y, s) is generated satisfying

$$s^T x \leq \varepsilon_0, \quad \|Ax - b\| \leq \varepsilon_P, \quad \text{and } \|A^T y + s - c\| \leq \varepsilon_D \qquad (7.32)$$

(an approximately optimal point; more precisely, x and (y, s) are ε_0-optimal in the nearby problems where b is replaced by Ax and c by $A^T y + s$), or

$$\|(x, s)\|_1 > \omega^*, \qquad (7.33)$$

for suitable positive (small) ε_0, ε_P, ε_D and (large) ω^*. KMM argue (§4 of [10]) that, in the latter case, there is no feasible solution in a large region of $\mathbb{R}_{++}^n \times \mathbb{R}^m \times \mathbb{R}_{++}^n$. A slight modification of their algorithm (§5 of [10]) yields stronger conclusions, but neither version appears to generate a certificate of infeasibility.

KMM prove (§4 of [10]) that for given positive ε_0, ε_P, ε_D and ω^*, their algorithm terminates finitely. We now show how their method can provide approximate certificates of infeasibility. Suppose that (P) is strictly infeasible, and that ε_P is chosen sufficiently small that there is no nonnegative solution to $\|Ax - b\| \leq \varepsilon_P$ (see Proposition 7.1).

Theorem 7.8 *Suppose the KMM algorithm is applied to a primal strictly infeasible instance, with ε_P chosen as above and the large norm termination criterion (7.33) disabled. Then $\|s_k\| \to \infty$, $\beta_k := b^T y_k \to \infty$ and there is a subsequence along which $(y_k/\beta_k, s_k/\beta_k) \to (\bar{y}, \bar{s})$, with the latter a certificate of primal infeasibility.*

Proof By our choice of ε_P, the algorithm cannot terminate due to (7.32), and we have disabled the other termination criterion, so that the method generates an infinite sequence of iterates (x_k, y_k, s_k). KMM show that, if $\|(x_k, s_k)\|_1 \leq \omega$, for any positive ω, then there is some $\underline{\alpha} > 0$, depending on ω, such that $\bar{\alpha}_k \geq \underline{\alpha}$, and hence, by (7.31), the total complementarity $s^T x$ decreases at least by the factor $[1 - \underline{\alpha}(1 - \sigma_3)] < 1$ at this iteration. On every iteration, the total complementarity does not increase. Hence, if there is an infinite number of iterations with $\|(x_k, s_k)\|_1 \leq \omega$, $s_k^T x_k$ converges to zero, and since all iterates lie in $\mathcal{N}$, $\|Ax_k - b\|$ also tends to zero. But this contradicts strict primal infeasibility, so that there cannot be such an infinite subsequence. This holds for any positive ω, and thus $\|(x_k, s_k)\| \to \infty$. By Proposition 7.1, $\{x_k\}$ remains bounded, so $\|s_k\| \to \infty$. By the rules of the KMM algorithm, $s_k^T x_k \leq s_0^T x_0$ for all k. Hence by Proposition 7.2, $\beta_k \to \infty$. From (7.29) we see that $(s_k/\beta_k)^T x_0$ and thus $\bar{s}_k := s_k/\beta_k$ remain bounded, so that there is a infinite subsequence K with $\lim_K \bar{s}_k := \lim_{k \in K, k \to \infty} \bar{s}_k = \bar{s}$ for some $\bar{s} \geq 0$. Further, $\bar{y}_k := y_k/\beta_k$ satisfies $A^T \bar{y}_k = c_k/\beta_k - \bar{s}_k$, which converges to $-\bar{s}$ along K, since c_k remains bounded. Hence $\bar{y}_k$ converges to $\bar{y} := -(AA^T)^{-1} A\bar{s}$ along this subsequence. We therefore have

$$A^T \bar{y} + \bar{s} = \lim_K (A^T y_k + s_k)/\beta_k$$
$$= \lim_K c_k/\beta_k = 0, \quad \bar{s} \geq 0, \quad b^T \bar{y} = \lim_K b^T y_k/\beta_k = 1,$$

as desired. □

While an exact certificate of infeasibility is obtained only in the limit (except under the happy circumstance that $A^T \bar{y}_k \leq 0$ for some k),

$(\bar{y}_k, \bar{s}_k)$ is an approximate such certificate for large k, and we can conclude that there is no feasible x in a large region, and that a nearby problem with slightly perturbed A matrix is primal infeasible; see Todd and Ye [28].

The results above shed light on our assumptions in §7.4.2. Indeed, we showed that $b^T y_k \to \infty$, which justifies our supposition that $\beta > 0$ in Assumption 7.1. As we noted in §7.4.2, Assumption 7.2 (or equivalently 7.3) holds if σ (or $\bar{\sigma}$) is sufficiently small (depending on the current iterate), although this may contradict the KMM choice of σ. In practice, even with empirical rules for choosing the parameters, the assumptions that $\beta > 0$ and $\Delta\beta > 0$ seem to hold after the first few iterations. The main assumption left is that (y, s) is feasible, and we have not been able to establish rules for choosing α_P and α_D that will assure this (it is necessary to have $\alpha_D = 1$ at some iteration, unless (y_0, s_0) is itself feasible). As we noted, this assumption does seem to hold in practice. Moreover, if $A^T y_k + s_k$ converges to c but never equals it, then eventually $\|A^T y_k + s_k - c\| \leq \varepsilon_D$, and then KMM's modified algorithm (§5 of [10]) replaces c by $c_k = A^T y_k + s_k$, so that the dual iterates are from now on feasible in the perturbed problem.

Finally, let us relate the neighborhood conditions for an iterate in the 'real' universe to those for the corresponding shadow iterate. Let us suppose that the current iterate (x, y, s) satisfies Assumption 7.1, and let $(\bar{x}, \bar{\zeta}, \bar{y}, \bar{s})$ be the corresponding shadow iterate. We define the neighborhood $\bar{\mathcal{N}}$ in the shadow universe using parameters $\bar{\gamma}_0, \bar{\gamma}_P, \bar{\gamma}_D$, $\bar{\varepsilon}_P$, and $\bar{\varepsilon}_D$ in the obvious way, with the centering condition involving only the $\bar{x}$- and $\bar{s}$-variables, since $\bar{\zeta}$ is free.

Proposition 7.3 *Suppose $\varepsilon_P \leq s^T x/\gamma_P$ and $\bar{\varepsilon}_D \leq \bar{s}^T \bar{x}/\bar{\gamma}_D$. Then, if $\bar{\gamma}_0 = \gamma_0$ and $\bar{\gamma}_D = (\|Ax_0 - b\|/\|c\|)\gamma_P$, $(x, y, s) \in \mathcal{N}$ if and only if $(\bar{x}, \bar{\zeta}, \bar{y}, \bar{s}) \in \bar{\mathcal{N}}$.*

(Note that our requirements on the γ's are natural; γ_0 and $\bar{\gamma}_0$ are dimension-free, while we expect γ_P to be inversely proportional to a typical norm for $Ax - b$, such as $\|Ax_0 - b\|$, and $\bar{\gamma}_D$ to be inversely proportional to a typical norm for $A^T \bar{y} + \bar{s} - 0$, such as $\|c\|$.)

Proof Since $\bar{x} = x/\varphi$ and $\bar{s} = s/\beta$, we have $\bar{s}^T \bar{x} = s^T x/(\beta\varphi)$ and $\bar{\mu} = \mu/(\beta\varphi)$ where $\mu := s^T x/n$ and $\bar{\mu} := \bar{s}^T \bar{x}/n$. Thus, for each j,

$$\bar{s}_j \bar{x}_j \geq \bar{\gamma}_0 \bar{\mu} \text{ iff } s_j x_j/(\beta\varphi) \geq \bar{\gamma}_0 \mu/(\beta\varphi) \text{ iff } s_j x_j \geq \gamma_0 \mu.$$

Next, $\|A^T y + s - c\| = 0 \leq \max(\varepsilon_D, s^T x/\gamma_D)$ and $\|A\bar{x} + \bar{\zeta} - Ax_0\| = 0 \leq \max(\bar{\varepsilon}_P, \bar{s}^T \bar{x}/\bar{\gamma}_P)$. Finally, $Ax - b = \varphi(Ax_0 - b)$, so

$$\varphi\|Ax_0 - b\| = \|Ax - b\| \leq \max(\varepsilon_P, s^T x/\gamma_P) = s^T x/\gamma_P$$

if and only if

$$\varphi \leq s^T x/(\gamma_P \|Ax_0 - b\|);$$

whereas $b^T \bar{y} - 1 = 0$ and $A^T \bar{y} + \bar{s} - 0 = c/\beta$, so

$$\|c\|/\beta = \|A^T \bar{y} + \bar{s} - 0\| \leq \max(\bar{\varepsilon}_D, \bar{s}^T \bar{x}/\bar{\gamma}_D) = \bar{s}^T \bar{x}/\bar{\gamma}_D = s^T x/(\beta\varphi\bar{\gamma}_D)$$

if and only if

$$\varphi \leq s^T x/(\bar{\gamma}_D \|c\|).$$

By our conditions on γ_P and $\bar{\gamma}_D$, these conditions are equivalent. $\square$

Let us summarize what we have shown (and not shown) about the convergence of IIP methods. (Of course, analogous results for the dual strictly infeasible case can easily be established.) Theorem 7.8 shows that the original KMM algorithm will provide certificates of infeasibility in the limit for strictly infeasible instances. However, our development of §7.4.2 and §7.4.3 suggests a more ambitious goal. We would like a strategy for choosing the centering parameter σ and the step sizes α_P and α_D at each iteration so that:

(a) In case (P) and (D) are feasible, the iterates converge to optimal solutions to these problems;

(b) In case (P) is strictly infeasible, the iterates become dual feasible, $b^T y$ becomes positive, and thenceforth the shadow iterates converge to optimal solutions of $(\bar{P})$ and $(\bar{D})$, unless a certificate of primal infeasibility is generated;

(c) In case (D) is strictly infeasible, the iterates become primal feasible, $c^T x$ becomes negative, and thenceforth the shadow iterates converge to optimal solutions of $(\tilde{P})$ and $(\tilde{D})$, unless a certificate of dual infeasibility is generated.

Of course, the algorithm should proceed without knowing which case obtains. We would further like some sort of polynomial bound on the number of iterations required in each case.

Unfortunately, we are a long way from achieving this goal. We do not know how to achieve dual (primal) feasibility in case (b) (case (c)). And we do not know how to choose the parameter σ and corresponding $\bar{\sigma}$

and the step sizes α_P and α_D and corresponding $\bar{\alpha}_P$ and $\bar{\alpha}_D$ to achieve simultaneously good progress in (P) and (D) and $(\bar{P})$ and $(\bar{D})$ (which we would like since we do not know which of cases (a) and (b) holds). The next subsection gives some practical guidance for choosing these parameters, without any guarantees of convergence.

7.5.3 Implications for the design of IIP methods

Suppose we are at a particular iteration of an IIP method and we suspect that (P) is strictly infeasible. (Similar considerations of course apply in the dual case.) For example, we might check that φ (the proportion of primal infeasibility remaining at the current iterate x) is at least .01 whereas the dual iterate (y, s) is (approximately) feasible, and $\beta := b^T y$ and $\Delta\beta := b^T \Delta y$ are positive. It then might make practical sense to choose the parameters to determine the next iterates with some attention to the (presumably better behaved) shadow iterates. Let us recall the relationship between the parameters in the real and shadow universes:

$$\bar{\sigma} := \frac{\beta}{\Delta\beta}\sigma, \quad \bar{\alpha}_P := \frac{\alpha_P}{1-\alpha_P} \cdot \frac{\Delta\beta}{\beta}, \quad \bar{\alpha}_D := \frac{\alpha_D \Delta\beta}{\beta + \alpha_D \Delta\beta}.$$

In the case that we expect, $\Delta\beta$ will be considerably larger than β, so that $\bar{\sigma}$ will be much smaller than σ. Practical rules for choosing σ might lead to a value close to 1, since poor progress is being made in achieving feasibility in (P); but $\bar{\sigma}$ may still be quite small, indicating good progress toward optimality in $(\bar{P})$ and $(\bar{D})$. Indeed, it seems reasonable to choose σ quite large, so that $\bar{\sigma}$ is not too small — recall that merely achieving feasibility in $(\bar{D})$ yields a certificate of primal infeasibility; an optimal solution is not required. Of course, $\Delta\beta$ itself depends on σ by the relation above Assumption 7.2, but choosing a larger σ is likely to increase $\bar{\sigma}$.

Having thus chosen σ, we need to choose the step size parameters α_P and α_D. Because primal feasibility cannot be achieved, $\alpha_P < 1$, but again, the resulting $\bar{\alpha}_P$ may be much larger, indeed even bigger than 1. In such a case it seems reasonable to make α_P even smaller, so that the corresponding $\bar{\alpha}_P = 1$, using the formula above. A reverse situation occurs for α_D and $\bar{\alpha}_D$. If we limit α_D to 1, the corresponding $\bar{\alpha}_D$ may be quite small, whereas we would like to have $\bar{\alpha}_D = 1$ to obtain a certificate of infeasibility. Such a value corresponds to $\alpha_D = \infty$, so that it seems reasonable to take α_D as a large fraction of the distance to the boundary, even if this exceeds 1. If it is possible to choose $\alpha_D = \infty$, then $(\Delta y, \Delta s)$ is itself a certificate of primal infeasibility.

Modifications of this kind in the software package SDPT3 (see [31]) seem quite useful to detect infeasibility; in particular, allowing α_D (α_P) to be very large when primal (dual) infeasibility is suspected usually gives a very good certificate of primal (dual) infeasibility at the next iteration.

7.6 Extensions to conic programming

All of our discussion so far has concentrated on the linear programming case. In this section we show that the results of §7.4 extend to many IIP methods for more general conic programming problems of the form

$$(\check{P}) \text{ minimize } \langle c, x \rangle,$$
$$Ax = b, \quad x \in K.$$

Here K is a closed, convex, pointed (i.e., containing no line), and solid (with nonempty interior) cone in a finite-dimensional real vector space E with dual E^*, and $c \in E^*$: $\langle s, x \rangle$ denotes the result of $s \in E^*$ acting on $x \in E$. A is a surjective linear transformation from E to the dual Y^* of another finite-dimensional real vector space Y, and $b \in Y^*$. In particular, this includes the case of semidefinite programming (SDP), where $E = E^*$ is the space of symmetric matrices of order n with the inner product $\langle s, x \rangle := \text{Trace}(s^T x)$ and K is the positive semidefinite cone. It also contains the case of second-order cone programming (SOCP), where $E = E^* = \mathbb{R}^n$ with the usual inner product and $K = K_1 \times \cdots \times K_q$, with $K_i := \{x_i \in \mathbb{R}^{n_i} : x_i = (x_{i1}; \bar{x}_i), x_{i1} \geq \|\bar{x}_i\|\}$ and $\sum_i n_i = n$. (Here we have used Matlab notation: $(u; v)$ is the column vector obtained by concatenating the column vectors u and v. Hence the first component of x_i is required to be at least the Euclidean norm of the vector of the remaining components.)

Both of these classes of optimization problems have nice theory and wide-ranging applications: see, e.g., Ben-Tal and Nemirovski [2] or Todd [27].

The problem dual to ($\check{P}$) is

$$(\check{D}) \text{ maximize } \langle b, y \rangle,$$
$$A^* y + s = c, \quad s \in K^*,$$

where $A^* : Y \to E^*$ is the adjoint transformation to A and $K^* := \{s \in E^* : \langle s, x \rangle \geq 0 \text{ for all } x \in K\}$ is the cone dual to K. In the two cases above, K is self-dual, so that $K^* = K$ (we have identified E and E^*).

Given a possibly infeasible interior point $(x, y, s) \in \operatorname{int} K \times Y \times \operatorname{int} K^*$, a primal-dual IIP method (see, e.g., [19, 20, 21, 27, 2]) takes steps in the directions $(\Delta x, \Delta y, \Delta s)$ obtained from a linear system of the form

$$\begin{aligned} A^* \Delta y + \Delta s &= c - A^* y - s, \\ A \Delta x &= b - Ax, \\ \mathcal{E} \Delta x + \mathcal{F} \Delta s &= \sigma g - h, \end{aligned} \qquad (7.34)$$

for certain operators $\mathcal{E} : E \to V$ and $\mathcal{F} : E^* \to V$ (V is another real vector space of the same dimension as E) and certain $g, h \in V$, depending on the current iterates x and s, and for a certain parameter $\sigma \in [0, 1]$; compare with (7.10).

We are again interested in the case that $(\check{P})$ or $(\check{D})$ is infeasible, and again we concentrate on the primal case, the dual being similar. We note that a sufficient condition for primal infeasibility is the existence of $(\bar{y}, \bar{s}) \in Y \times E^*$ with

$$A^* \bar{y} + \bar{s} = 0, \quad \bar{s} \in K^*, \quad \langle b, \bar{y} \rangle = 1, \qquad (7.35)$$

but in the general nonpolyhedral case this is no longer necessary. We will say that $(\check{P})$ is strictly infeasible if there is such a certificate with $\bar{s} \in \operatorname{int} K^*$ (again, this implies that $(\check{D})$ is strictly feasible). Henceforth we suppose that $(\check{P})$ is strictly infeasible and that a sequence of iterations from the initial infeasible interior point (x_0, y_0, s_0) has led to the current iterate (x, y, s) where the analogue of Assumption 7.1 holds (the only change is that $\beta := \langle b, y \rangle$ is positive). We consider the Farkas-like problem

$$\begin{aligned} (\bar{D}) \max \; & \langle A x_0, \bar{y} \rangle \\ & A^* \bar{y} + \bar{s} = 0, \\ & \langle b, \bar{y} \rangle = 1, \\ & \bar{s} \in K^*, \end{aligned}$$

with dual

$$\begin{aligned} (\bar{P}) \min \; & \bar{\zeta} \\ & A\bar{x} + b\bar{\zeta} = A x_0, \\ & \bar{x} \in K. \end{aligned}$$

We define the shadow iterate $(\bar{x}, \bar{\zeta}, \bar{y}, \bar{s})$ of (x, y, s) exactly as in Definition 7.1. We will show that, assuming $\mathcal{E}$, $\mathcal{F}$, g, and h depend on x and s suitably, once again an iteration from (x, y, s) corresponds appropriately to a shadow iteration from $(\bar{x}, \bar{\zeta}, \bar{y}, \bar{s})$. We define $\bar{\mathcal{E}}$, $\bar{\mathcal{F}}$, $\bar{g}$, and $\bar{h}$

from the shadow iterate as their unbarred versions were defined from (x, y, s).

Since we are assuming (y, s) feasible, our directions $(\Delta x, \Delta y, \Delta s)$ solve (7.34) with the first right-hand side replaced by zero. We again assume that $\Delta\beta := \langle b, \Delta y\rangle$ is positive. Having chosen positive step sizes α_P and α_D to obtain the new iterate (x_+, y_+, s_+), we define $\bar\alpha_P$, $\Delta\bar x$, $\Delta\bar\zeta$, $\bar\alpha_D$, $\Delta\bar y$, and $\Delta\bar s$ exactly as in §7.4.2.

Theorem 7.9 *Let us suppose that $\mathcal{E}$, $\mathcal{F}$, g, and h and $\bar{\mathcal{E}}$, $\bar{\mathcal{F}}$, $\bar g$, and $\bar h$ are related in one of the following ways:*

(a) $\bar{\mathcal{E}} = \mathcal{E}/\beta$, $\bar{\mathcal{F}} = \mathcal{F}/\varphi$, $\bar g = g/(\beta\varphi)$, and $\bar h = h/(\beta\varphi)$;
(b) $\bar{\mathcal{E}} = \mathcal{E}$, $\bar{\mathcal{F}} = (\beta/\varphi)\mathcal{F}$, $\bar g = g/\varphi$, and $\bar h = h/\varphi$; or
(c) $\bar{\mathcal{E}} = (\varphi/\beta)\mathcal{E}$, $\bar{\mathcal{F}} = \mathcal{F}$, $\bar g = g/\beta$, and $\bar g = g/\beta$.

Suppose also that $\mathcal{E}x = \mathcal{F}s = h$. Then the directions $(\Delta\bar x, \Delta\bar\zeta, \Delta\bar y, \Delta\bar s)$ solve the Newton system for $(\bar P)$ and $(\bar D)$ given below:

$$\begin{aligned}
A^*\Delta\bar y + \Delta\bar s &= -A^*\bar y - \bar s,\\
\langle b, \Delta\bar y\rangle &= 0,\\
A\Delta\bar x + b\Delta\bar\zeta &= 0,\\
\bar{\mathcal{E}}\Delta\bar x + \bar{\mathcal{F}}\Delta\bar s &= \bar\sigma\bar g - \bar h,
\end{aligned} \qquad (7.36)$$

for the value $\bar\sigma := \frac{\beta}{\Delta\beta}\sigma$.

Proof The derivation is exactly as in the proof of Theorem 7.4 except for that of the last equation. In case (a) we obtain

$$\begin{aligned}
\bar{\mathcal{E}}\Delta\bar x + \bar{\mathcal{F}}\Delta\bar s &= \frac{1}{\beta}\mathcal{E}\left(\frac{\beta}{\varphi\Delta\beta}(\Delta x + x)\right) + \frac{1}{\varphi}\mathcal{F}\left(\frac{\Delta s}{\Delta\beta} - \frac{s}{\beta}\right)\\
&= \frac{1}{\varphi\Delta\beta}(\mathcal{E}\Delta x + \mathcal{F}\Delta s) + \frac{1}{\varphi\Delta\beta}\mathcal{E}x - \frac{1}{\beta\varphi}\mathcal{F}s\\
&= \frac{1}{\varphi\Delta\beta}(\sigma g - h) + \frac{1}{\varphi\Delta\beta}h - \frac{1}{\beta\varphi}h\\
&= \left(\frac{\beta}{\Delta\beta}\sigma\right)\left(\frac{g}{\beta\varphi}\right) - \left(\frac{h}{\beta\varphi}\right) = \bar\sigma\bar g - \bar h,
\end{aligned}$$

as desired. Cases (b) and (c) are exactly the same after dividing the last equation by β (case (b)) or φ (case (c)). $\square$

Note that case (a) covers any situation where $\mathcal{E}$ scales with s, $\mathcal{F}$ with x, and g and h with both x and s. (As long as we also have $\mathcal{E}x = \mathcal{F}s = h$.) This includes our previous linear programming analysis, where $\mathcal{E} = S$,

$\mathcal{F} = X$, $\gamma = \mu e$, and $h = SXe$. It also includes the Alizadeh-Haeberly-Overton [1] direction for SDP, where $\mathcal{E}$ is the operator $v \to (sv + vs)/2$, $\mathcal{F}$ the operator $v \to (vx + xv)/2$, $g \langle s, x \rangle / n$ times the identity, and $h = (sx + xs)/2$. (We write direction instead of method here and below, to stress that we are concerned here with the Newton system, which defines the direction; many different methods can use this direction, depending on their choices of the centering parameter and the step sizes.)

As a second example, the HRVW/KSH/M direction for SDP (see [8, 11, 17]) has $\mathcal{E}$ the identity, $\mathcal{F}$ the operator $v \to (xvs^{-1} + s^{-1}vx)/2$, $g \langle s, x \rangle / n$ times s^{-1}, and $h = x$. It is easily seen that these choices satisfy the conditions of case (b), as well as the extra condition. Another instance of case (b) is the Nesterov-Todd (NT) direction for SDP — see [20, 21]. Here $\mathcal{F}$ is the operator $v \to wvw$, where $w := x^{1/2}[x^{1/2}sx^{1/2}]^{-1/2}x^{1/2}$ is the unique positive definite matrix with $wsw = x$, and $\mathcal{E}$, g, and h are as above. Then, if $\bar{w}\bar{s}\bar{w} = \bar{x}$, it is easy to see that $\bar{w} = (\beta/\varphi)^{1/2}w$, so again the conditions are simple to check.

The dual HRVW/KSH/M direction for SDP (see [11, 17]) is an instance of case (c). Here $\mathcal{E}$ takes v to $(svx^{-1} + x^{-1}vs)/2$, $\mathcal{F}$ is the identity, $g \langle s, x \rangle / n$ times x^{-1}, and $h = s$.

We presented the NT direction above in the form that is most useful for computing the directions, and only for SDP. But it is applicable in more general self-scaled conic programming (including SOCP), using a self-scaled barrier function, and can be given in a form as above satisfying the conditions of case (b), or another form that conforms to case (c).

Lastly, the presentations of the HRVW/KSH/M and NT directions for SOCP in Tsuchiya [29] use different forms of the Newton system: and it is easy to see that these fit into case (a) of the theorem.

Let us finally note that our results on boundedness and unboundedness in §7.5.1 also hold for general conic programming problems. The key simple fact is that, if $s \in \text{int } K^*$, then $\{x \in K : \langle s, x \rangle \leq \delta\}$ is bounded for any positive δ. Hence analogues of Propositions 7.1 and 7.2 hold.

7.7 Concluding remarks

We have shown that there is a surprising connection between the iterates of an IIP method, applied to a dual pair of problems (P) and (D) in the case that one of them is strictly infeasible, and those of another IIP method applied to a related pair of strictly feasible problems whose optimal solutions give a certificate of infeasibility for (P) or (D). This connection involves a projective transformation from the original setting of the problems to a 'shadow' universe, where the corresponding iterates lie. It holds not only for linear programming, but also for a range of methods for certain more general conic programming problems, including semidefinite and second-order cone programming problems. We hope that an intriguing glimpse of this connection has been provided, but it is clear that much work remains to be done to understand the convergence of IIP methods.

Acknowledgment

The author would like to thank Bharath Rangarajan for helpful discussions; in particular, he improved my earlier analysis of the Kojima-Megiddo-Mizuno algorithm, in which I only proved that the limit superior of the dual objective functions $b^T y_k$ was infinite, not that the whole sequence diverged.

References

[1] Alizadeh, F., Haeberly, J.-P. A. and Overton, M.L. (1998). Primal-dual interior-point methods for semidefinite programming: convergence rates, stability and numerical results, *SIAM J. Optim.* **8**, 746–768.
[2] Ben-Tal, A. and Nemirovski, A.S. (2001). *Lectures on Modern Convex Optimization: Analysis, Algorithms, and Engineering Applications* (SIAM Publications. SIAM, Philadelphia, USA).
[3] Bixby, R.E. (2001). Solving real-world linear programs: a decade and more of progress. *Oper. Res.* **50**, 3–15.
[4] Dantzig, G.B. (1963). *Linear Programming and Extensions* (Princeton University Press. Princeton).
[5] Forsgren, A., Gill, P.E. and Wright, M.H. (2002). Interior methods for nonlinear optimization. *SIAM Rev.* **44**, 525–597.
[6] Goldman, A.J. and Tucker, A.W. (1956). Theory of linear programming. In Kuhn, H.W. and Tucker, A.W., editors, *Linear Equalities and Related Systems* (Princeton University Press, Princeton, N. J.), 53–97.
[7] Güler, O. and Ye, Y. (1993). Convergence behavior of interior point algorithms, *Math. Progr.* **60**, 215–228.
[8] Helmberg, C., Rendl, F., Vanderbei, R. and Wolkowicz, H. (1996). An interior-point method for semidefinite programming, *SIAM J. Optim.* **6**, 342–361.
[9] Karmarkar, N.K. (1984). A new polynomial-time algorithm for linear programming, *Combinatorica* **4**, 373–395.
[10] Kojima, M., Megiddo, N. and Mizuno, S. (1993). A primal–dual infeasible–interior–point algorithm for linear programming, *Math. Progr.* **61**, 263–280.

[11] Kojima, M., Shindoh, S. and Hara, S. (1997). Interior-point methods for the monotone semidefinite linear complementarity problem in symmetric matrices, *SIAM J. Optim.* **7**, 86–125.

[12] Lustig, I.J. (1990/91). Feasibility issues in a primal–dual interior point method for linear programming, *Math. Progr.* **49**, 45–162.

[13] Lustig, I.J., Marsten, R.E. and Shanno, D.F. (1991). Computational experience with a primal–dual interior point method for linear programming, *Lin. Alg. Appl.* **152**, 191–222.

[14] Mizuno, S. (1994). Polynomiality of infeasible–interior–point algorithms for linear programming, *Math. Progr.* **67**, 109–119.

[15] Mizuno, S. and Todd, M.J. (2001). On two homogeneous self-dual systems for linear programming and its extensions, *Math. Progr.* **89**, 517–534.

[16] Mizuno, S., Todd, M.J. and Tunçel, L. (1994). Monotonicity of primal and dual objective values in primal–dual interior–point algorithms, *SIAM J. Optim.* **4**, 613–625.

[17] Monteiro, R.D.C. (1997). Primal-dual path-following algorithms for semidefinite programming, *SIAM J. Optim.* **7**, 663–678.

[18] Nemhauser, G.L., Rinnooy Kan, A.H.G. and Todd, M.J., editors (1989). *Optimization*, volume 1 of *Handbooks in Operations Research and Management Science* (North Holland. Amsterdam, The Netherlands).

[19] Nesterov, Y.E. and Nemirovskii, A.S. (1994). *Interior Point Polynomial Methods in Convex Programming: Theory and Algorithms* (SIAM Publications. SIAM, Philadelphia, USA).

[20] Nesterov, Y.E. and Todd, M.J. (1997). Self-scaled barriers and interior-point methods for convex programming, *Math. Oper. Res.* **22**, 1–42.

[21] Nesterov, Y.E. and Todd, M.J. (1998). Primal-dual interior-point methods for self-scaled cones, *SIAM J. Optim.* **8**, 324–364.

[22] Nesterov, Y.E., Todd, M.J. and Ye, Y. (1999). Infeasible-start primal-dual methods and infeasibility detectors for nonlinear programming problems, *Math. Progr.* **84**, 227–267.

[23] Potra, F.A. (1996). An infeasible interior–point predictor–corrector algorithm for linear programming, *SIAM J. Optim.* **6**, 19–32.

[24] Renegar, J. (2001). *A Mathematical View of Interior-Point Methods in Convex Optimization* (SIAM Publications. SIAM, Philadelphia, USA).

[25] Schrijver, A. (1986). *Theory of Linear and Integer Programming* (John Wiley & Sons, New York).

[26] Todd, M.J. (2002). The many facets of linear programming, *Math. Progr.* **91**, 417–436.

[27] Todd, M.J. (2001). Semidefinite optimization, *Acta Numer.* **10**, 515–560.

[28] Todd, M.J. and Ye, Y. (1998). Approximate Farkas lemmas and stopping rules for iterative infeasible-point algorithms for linear programming, *Math. Progr.* **81**, 1–21.

[29] Tsuchiya, T. (1999). A convergence analysis of the scaling-invariant primal-dual path-following algorithms for second-order cone programming, *Optim. Meth. Soft.* **11/12**, 141–182.

[30] Tütüncü, R.H. (1999). An infeasible-interior-point potential-reduction algorithm for linear programming, *Math. Progr.* **86**, 313–334.

[31] Tütüncü, R.H., Toh, K.C. and Todd, M.J. (2003). Solving semidefinite-quadratic-linear programs using SDPT3, *Math. Progr.*.

[32] Wright, M.H. (1992). Interior methods for constrained optimization, *Acta Numer.* **1**, 341–407.

[33] Wright, S.J. (1997). *Primal-Dual Interior-Point Methods* (SIAM, Philadelphia).

[34] Xu, X., Hung, P.F. and Ye, Y. (1996). A simplified homogeneous and self-dual linear programming algorithm and its implementation, *Ann. Oper. Res.* **62**, 151–171.

[35] Ye, Y., Todd, M.J. and Mizuno, S. (1994). An $\mathcal{O}(\sqrt{n}L)$-iteration homogeneous and self-dual linear programming algorithm *Math. Oper. Res.* **19**, 53–67.

[36] Zhang, Y. (1994). On the convergence of a class of infeasible interior-point methods for the horizontal linear complementarity problem, *SIAM J. Optim.* **4**, 208–227.

8

Maple Packages and Java Applets for Classification Problems in Geometry and Algebra

Ian M. Anderson
Department of Mathematics and Statistics
Utah State University
Logan, Utah 84322-3900, USA
Email: anderson@math.usu.edu

Abstract

In this note I shall briefly describe an on-going project at Utah State University to create Maple software and Java applets for use in classification problems in geometry and algebra. The first problem we are currently working on deals with Petrov's remarkable classification of all 4-dimensional local group actions which admit a Lorentz invariant metric; the second focuses on the classification of low dimensional Lie algebras. The software which supports our work on these two classification problems is part of the Maple suite of software packages called *Vessiot*, an integrated collection of Maple programs for computations in differential geometry, Lie algebras and the variational calculus on jet spaces.

8.1 Introduction

A central theme in mathematics is the classification of mathematical objects up to a prescribed notion of equivalence. Well-known examples of such classification problems include the classification of two dimensional surfaces up to homeomorphism, the classification of Abelian groups up to isomorphism, and the classification of vector field systems in the plane up to local diffeomorphism.

From a strictly mathematical viewpoint, a classification problem may be considered solved once a complete list of inequivalent representatives of the mathematical objects under classification is obtained. However, from a practical perspective, a number of important issues remain.

- How does one locate a given mathematical object in the list of representatives furnished by the solution of the classification problem? This problem is especially acute if one wants to use the results of a classification problem but one is not an expert in the area.
- Once one has located a given object in the list of representatives, how does one find the transformation which maps the given object to its representative?
- How can one search through the list of representatives furnished by the solution of the classification problem to identify those representatives with a prescribed set of properties.

The purpose of this short article is to describe some of the on-going work with students and colleagues at Utah State University to create websites and data-based Java applets to address some of these practical issues regarding classification problems in mathematics. I also wish to introduce the Maple software suite *Vessiot*, which we have used to perform the extensive computations needed to create the data-bases for these websites.

The original HTML files for my lecture given during the FoCM conference at the University of Minnesota, upon which these notes are based, are available at *www.math.usu.edu/~anderson/FoCM_talk*. *Vessiot*, together with accompanying help files and tutorials, can be downloaded from the Symbolics page of the Formal Geometry and Mathematical Physics website *www.math.usu.edu/~fg_mp*. This page contains a link to the *Guided Tour* worksheet for *Vessiot* which provides a good overview of the many capabilities of this software.

The websites for the Petrov classification of spacetimes and for the classification of Lie algebras are located at *www.matti.usu.edu/Petrov* and *www.matti.usu.edu/Lie*. I encourage the reader to visit these two sites and to run the Java applets contained within. While still under construction, these applets nevertheless demonstrate the great potential of the web for disseminating the solutions to various mathematical classification problems.

8.2 Vessiot

The Maple package *Vessiot* was initially conceived to provide symbolic software for the calculus of vector fields and differential forms on jet spaces and to implement all the various operations and procedures in the variational bicomplex [1]. From the outset the primary motivation

8. Maple Packages and Java Applets

in developing this software was to create a general purpose package for computations in differential geometry, with an emphasis on the geometry of differential equations, which would support development of a wide range of specialized applications.

- *Vessiot* contains an extensive set of commands for creating vector fields, differential forms, tensors, transformations and so on and for the manipulation of these objects. Mathematical objects can be created using a simple and natural syntax with the Vessiot parsing command and are displayed in an easy-to-read mathematical format.
- *Vessiot* has wide capabilities for working with multiple coordinates systems simultaneous within a single Maple session and for constructing transformations between such coordinates. This capability is invaluable in the transformation theory of differential equations, in the theory of moving frames and for the classification of low dimensional Lie algebras.
- The *Vessiot* environment is integrated – many commands work across a wide range of contexts. The internal data structures upon which *Vessiot* is constructed are used to properly determine the correct context for such commands.
- *Vessiot* contains a wide range of programming utilities which enable the user to quickly write new applications.
- *Vessiot* contains extensive libraries of Lie algebras, vector field systems, Lorentz metrics, differential equations and their symmetries and so on, all of which are immediately available for use.
- extensive, detailed help files are available through the Maple help command.

The main package of the *Vessiot* suite contains all the basic commands for the calculus on jet spaces. These include:

- commands to create coordinate systems of independent and dependent variables and to manipulate the associated jet spaces.
- a wide variety of commands for inputting vector fields, differential forms, contact forms, transformations and differential equations.
- arithmetic operations for addition, scalar multiplication, and interior and wedge products of differential forms and vector fields.
- the differential operations of exterior differentiation of forms, horizontal and vertical differentiation of biforms, and Euler operators.
- the homotopy operators for the deRham complex and the variational bicomplex.

- commands for the composition and inverses of transformations.
- prolongation of vector fields, transformations and differential equations.
- frame procedures for performing calculations with frames other than the coordinate frame.
- extensive linear algebra utilities for working with vector fields and differential forms.
- *Vessiot* programming utilities for accessing data pertaining to coordinate systems and for accessing various parts of the internal representation of all Vessiot objects.

On top of this main package a host of more specialized packages have been developed. These include

de_appls: a differential equations package which includes routines for Noether's theorem and for the symmetry reduction of differential equations.
eds: a rudimentary exterior differential systems package.
Frobenius: a rudimentary package for solving Frobenius systems of differential equations.
Gelfand_Dickey: a package of commands for the Gelfand Dickey transform.
group_actions: computations with group actions on manifolds.
invariant_metrics: a package for computing curvature tensors for invariant metrics on homogeneous spaces.
isometries: calculates the Lie algebra of isometries of a metric.
Koszul: an extensive package of commands for general Lie algebra computations.
moving_frame: a package for constructing moving frames and differential invariants on jet spaces.
Mubar: a package for the classification of low dimensional Lie algebras.
Spencer: a package for tableaux computations and Spence cohomology.
tensors: an extended tensor package which allows for manipulation of tensors on jet spaces.

The last section of this article contains a few simple illustrations of *Vessiot*. They are, admittedly, a poor substitute for downloading the program and experimenting with it.

Let me conclude this brief overview by acknowledging the many students at Utah State University who have made essential contributions to the development of *Vessiot* over the years. Bryan Croft created the origin package in Macsyma; Cinnamon Hillyard converted Bryan's package to Maple and introduced many of the current data structures; Charles Miller added multiple coordinate functionality, the user friendly output routines, and the tensors package; Jeff Humphries, Florin Cantrina, Jamie Jorgenson and Robert Berry made essential contributions to the Lie algebra package; and the class of Math 5820 played a critical role in the development of the Lie algebra classification package. Many thanks also to my colleagues Mark Fels and Charles Torre for their constant assistance with software testing and debugging.

8.3 Petrov on line

In the book Einstein Spaces, Petrov [5] gives a complete list of all infinitesimal group actions on a 4-dimensional manifold which preserve a Lorentz metric. Our initial interest in Petrov's work was as a source of interesting group actions to illustrate our recent theoretical work on group invariant solutions [3] and the principle of symmetric criticality [2].

However, it is difficult to work systematically with Petrov's classification and it is almost impossible to positively identify even the standard well-known metrics within Petrov's list of metrics. With *Vessiot*, we were able to easily compute many invariant properties for the group actions in Petrov's book and this suggested the possibility of constructing a database of such properties which would allow one to systematically identify a given metric within the Petrov classification.

Rather that describe here in detail the Petrov On Line website, it is much easier to simply point the interested reader to the aforementioned address. To run the Petrov on Line applet your browser will need the current Java plugin – this will automatically be installed if necessary. Once the applet is running, click first on the reset button and then on the retrieve button. This returns the list of all the metrics in Petrov's book and then, by highlighting a particular metric, its properties are displayed in the various fields. Click on reset, enter in a few properties, and then click on retrieve to obtain a list of all metrics with the given properties.

The two *Vessiot* tutorials *Petrov On Line* and *isometries_tutorial*, which can be downloaded from the Symbolics page of the Formal Geometry and Mathematical Physics website (*www.math.usu.edu/~fg_mp*), illustrate the use of *Vessiot* to calculate the properties of a given metric required by Petrov On Line in order to locate the metric in Petrov's classification.

8.4 Lie algebras on line

Unlike the case of semi-simple Lie algebras, a good structure theory for solvable and nilpotent Lie algebras is unavailable. At this time the one alternative seems simply to construct extensive tables of these algebras. Accordingly, this appears to be another area where data-base applets could provide a useful tool in classification.

A major difficulty in the classification of solvable Lie algebras arises from the existence of continuous families of non-isomorphic solvable algebras. For example, consider the class of $n + 1$ dimensional Lie algebras with an n dimensional Abelian ideal. Such algebras are, roughly speaking, in one-to-one correspondence with the number of inequivalent Jordon canonical forms for $n \times n$ matrices. This difficulty is compounded by the fact that many basic numerical invariants of Lie algebras, such as the dimension of the derived algebra and dimension of the automorphism algebra, can change as the parameters in the structure constants for these Lie algebras vary. One can simply ignore this fact and be content with a very coarse classification or one can develop finer classifications by demanding that various invariants are constant within an allowable range of parameters. We are currently completing rather refined tables of all Lie algebras of dimensions ≤ 6 for which the dimensions of the derived, upper and lower series, the dimension of the automorphism algebra, and the dimensions of the cohomology spaces are constant.

The Lie Algebras on Line website provides a link to an extensive data-base of properties of all Lie algebras of dimensions 3 and 4. We are currently working to extend this data-base to algebras of dimensions 5 and 6.

8.5 Vessiot worksheets

8.5.1 Worksheet 1

In this worksheet the *Vessiot* user interface is illustrated. A coordinate system, *euc*, with three variables $[x, y, z]$ is defined, a few vector fields, differential forms and tensors are created, and some elementary calculations performed.

```
>    with(Vessiot):
>    coord_init([x,y,z],[],E3);
```

frame name : $E3$

8. Maple Packages and Java Applets 199

Lot of things happen in the background when the coord_init command is executed. In particular, the coordinate vector fields D_x, D_y and D_z and one forms dx, dy, dz are automatically defined. These are default labels for the coordinate vector fields and differential forms. The coord_init command allows one assign custom names to these vector fields and forms, if need be.

Define a couple of vector fields.

```
> X:= evalV(x*D_x - y*D_y);
```
$$X := x\, D\_x \; - y\, D\_y$$
```
> Y:= evalV(x^2*D_z);
```
$$Y := x^2\, D\_z$$

Compute their Lie bracket.

```
> Lie_bracket(X,Y);
```
$$2\, x^2\, D\_z$$

Define a couple of 1-forms.

```
> alpha:= evalV( dx -z*dy);
```
$$\alpha := dx \; - z\, dy$$
```
> beta:=evalV( dx +z*dz);
```
$$\beta := dx \; + z\, dz$$

Evaluate the interior product of a vector and a 1-form.

```
> hook(X,alpha);
```
$$z\, y + x$$

Compute the wedge product of a pair of forms.

```
> alpha &wedge beta;
```
$$z\, dx \,\hat{}\, dy \; + z\, dx \,\hat{}\, dz \; - z^2\, dy \,\hat{}\, dz$$

Compute the Lie derivative of a form.

> `Lie_derivative(Y,beta);`

$$2\,z\,x\,dx + x^2\,dz$$

Compute the exterior derivative of a form.

> `ext_d(alpha);`

$$dy \,\hat{}\, dz$$

Construct a tensor.

> `with(tensors):`
> `T:=evalV(dx &t D_z &t D_y);`

$$T := dx \; D_{\text{-}z} \; D_{\text{-}y}$$

Compute the Lie derivative of a tensor

> `Lie_derivative(X,T);`

$$2\,dx \; D_{\text{-}z} \; D_{\text{-}y}$$

Define another coordinate system, called S3, and a transformation from S3 to E3. This transformation defines S3 as the usual system of spherical coordinates in 3-dimensions.

> `coord_init([rho, theta,phi],[],S3);`

$$\textit{frame name} : S3$$

> `Phi:=transform(S3,E3,[x=rho*cos(theta)*sin(phi), y=rho*sin(theta)*sin(phi), z=rho*cos(phi)]);`

$$\Phi := [x = \varrho \cos(\theta)\sin(\varphi),\, y = \varrho\sin(\theta)\sin(\varphi),\, z = \varrho\cos(\varphi)]$$

Evaluate $x^2 + y^2 + z^2$ in spherical coordinates.

> `pullback(Phi, x^2+y^2+z^2);`

$$\varrho^2$$

Evaluate the 1-form dx in spherical coordinates.

> `pullback(Phi, dx);`

$$-\varrho \sin(\theta) \sin(\varphi) \, dtheta + \varrho \cos(\theta) \cos(\varphi) \, dphi$$

8.5.2 Worksheet 2

We define the Lagrangian $L = u_x^2 + u_y^2$ for Laplace's equation $u_{xx} + u_{yy} = 0$ in two independent variables. We check that this Lagrangian is rotational invariant and then use Noether's theorem to construct the associated conservation law.

This worksheet gives a glimpse of *Vessiot*'s capabilities for computations in jet spaces, the calculus of variations and the geometry of differential equations.

The default Vessiot notation for derivatives is $u_x = u[1, 0]$, $u_y = u[0, 1]$, $u_{xx} = u[2, 0]$, $u_{xy} = u[1, 1]$, $u_{yy} = u[0, 2]$ and so on. Alternative notation can be invoked.

> `with(Vessiot):`
> `coord_init([x,y],[u],E);`

$$frame\ name:\ E$$

Define the Lagrangian as a function.

> `L:= u[1,0]^2 + u[0,1]^2;`
$$L := u_{1,0}{}^2 + u_{0,1}{}^2$$

Define the Lagrangian as a form (or more precisely as a type (2,0) biform).

> `lambda:= evalV(L * Dx &w Dy);`

$$\lambda := (u_{1,0}{}^2 + u_{0,1}{}^2) \, Dx \,\hat{}\, Dy$$

Compute the Euler-Lagrange expression of L.

> `EL0(L);`
$$[-2\, u_{2,0} - 2\, u_{0,2}]$$

Define the one-parameter group of rotations in the xy plane.

```
> phi:=transform(E,E, [x=x*cos(theta) + y*sin(theta),
    y= -x*sin(theta) +y*cos(theta), u[0,0]=u[0,0]]);
```

$$\varphi := [x = x\cos(\theta) + y\sin(\theta),\ y = -x\sin(\theta) + y\cos(\theta),\ u_{0,0} = u_{0,0}]$$

Prolong the transformation φ to the 1-jets.

```
> phi1:=pr_transform(phi,1);
```

$$\varphi 1 := [x = x\cos(\theta) + y\sin(\theta),\ y = -x\sin(\theta) + y\cos(\theta),\ u_{0,0} = u_{0,0},$$
$$u_{1,0} = \cos(\theta)\, u_{1,0} + \sin(\theta)\, u_{0,1},\ u_{0,1} = -\sin(\theta)\, u_{1,0} + \cos(\theta)\, u_{0,1}]$$

To check the rotational invariance of the Lagrangian form λ, show that it remains unchanged under pullback by φ.

```
> pullback(phi1,lambda);
```

$$(u_{1,0}{}^2 + u_{0,1}{}^2)\, Dx \wedge Dy$$

To find the associated conservation law, first find the infinitesimal generator for the transformation φ.

```
> X:=transform_to_vect(phi,[theta])[1];
```

$$X := y\, D_{\_}x - x\, D_{\_}y$$

```
> with(de_appls);
```

$$[Lie\_reduction,\ Noether1,\ Noether2,\ inverse\_td]$$

```
> sigma:=Noether1(X,lambda);
```

$$\sigma := (-x\, u_{1,0}{}^2 + x\, u_{0,1}{}^2 - 2\, u_{0,1}\, u_{1,0}\, y)\, Dx$$
$$+ (y\, u_{1,0}{}^2 - y\, u_{0,1}{}^2 - 2\, u_{1,0}\, u_{0,1}\, x)\, Dy$$

To check that σ is a conservation law for Laplace's equation, take the horizontal exterior derivative of σ and note that the result vanishes on solutions to Laplace equations.

```
> check:=dH(sigma);
```

$$check := (2\, u_{2,0}\, u_{1,0}\, y - 2\, u_{2,0}\, u_{0,1}\, x - 2\, u_{0,2}\, u_{0,1}\, x + 2\, u_{0,2}\, u_{1,0}\, y)\, Dx \wedge Dy$$

```
> factor(check);
```

$$-2\left(u_{2,0} + u_{0,2}\right)\left(u_{0,1}\,x - u_{1,0}\,y\right) Dx \wedge Dy$$

To get a sense of some of the mathematics which Vessiot supports, let's look at the code for the command Noether1.

```
> op(Noether1);
```

> **proc**(*vector*, *Lagrangian*)
>   **local** ω, *boundary_form*, *Z*, *ans1*, *ans2*;
> ω := dV(*Lagrangian*) '&plus' EL_form(*Lagrangian*);
>   *boundary_form* := homotopy_dH(ω);
>   *Z* := pr_vect(evol(*vector*), objectOrder(*Lagrangian*));
>   *ans1* := vert_hook(*Z*, *boundary_form*);
>   *ans2* := total_hook(total(*vector*), *Lagrangian*);
>   *ans1* '&minus' *ans2*
>   **end**

The Noether formula for the conservation law defined by a Lagrangian λ and an infinitesimal symmetry X is

$$\sigma = -X_{\text{tot}} \lrcorner\, \lambda + \text{pr}X_{\text{ev}} \lrcorner\, \beta.$$

Here X_{tot} and X_{ev} are the total and evolutionary parts of the vector field X, $\text{pr}X_{\text{ev}}$ is the prolongation of X_{ev}. The form β is the boundary term in the first variational formula

$$d_V\lambda = E(\lambda) + d_H\beta,$$

where $E(\lambda)$ is the Euler-Lagrange form of λ. The first two lines of the code in the program *Noether1* compute the boundary form β; the next line prolongs the evolutionary part of the vector field X to the differential order of β; and the last two lines compute the two terms in the formula for the conserved form σ.

8.5.3 Worksheet 3

In this worksheet we shall demonstrate some capabilities of the Lie algebra package *Koszul*. This package is written on top of the basic *Vessiot* package and uses the same syntax and commands for manipulating vectors, form etc. We shall begin with the 6 dimensional Lie algebra of infinitesimal symmetries of the heat equation. These vectors are conveniently stored in the *Vessiot* library Olver_DE_lib which contains many of the worked problems from Olver's book [4].

```
> with(Vessiot):with(Vessiot_library):with(Koszul):
  with(Mubar):
> Gamma:=Olver_DE_lib([2,41])[3];
```

$\Gamma := [D\_x\ ,\ D\_t\ ,\ u_{0,0}\,D\_u_{[0,0]}\ ,\ 2t\,D\_t\ +x\,D\_x\ ,\ 2t\,D\_x$
$\qquad -x\,u_{0,0}\,D\_u_{[0,0]}\ ,4t^2\,D\_t\ +4t\,x\,D\_x\ -(x^2+2t)\,u_{0,0}\,D\_u_{[0,0]}\ ]$

The command vect_to_Lie_alg will compute the structure constants for the Lie algebra of vector fields Γ and return the result as a data structure used by Vessiot to initialize an abstract Lie algebra.

This is one of the significant design advantages of Vessiot – one came move effortlessly back and forth from the arena of differential geometry and vectors fields to that of Lie algebras.

```
> vect_to_Lie_alg(Gamma, Lie);
```

$[[Lie\_alg,\ Lie,\ [6]], [[[1,\ 4,\ 1],\ 1],\ [[1,\ 5,\ 3],\ -1],\ [[1,\ 6,\ 5],\ 2],\ [[2,\ 4,\ 2],\ 2],$
$\qquad [[2,\ 5,\ 1],\ 2],\ [[2,\ 6,\ 3],\ -2],\ [[2,\ 6,\ 4],\ 4],\ [[4,\ 5,\ 5],\ 1],\ [[4,\ 6,\ 6],\ 2]]]$

The command Lie_alg_init plays much the same role as the command coord_init. The names e1, e2, ... are assigned internal Vessiot representations as vectors in the Lie algebra and the structure equations are stored in memory.

```
> Lie_alg_init(%);
> Lie_bracket_mult_table();
```

	$e1$	$e2$	$e3$	$e4$	$e5$	$e6$
$e1$	0	0	0	$e1$	$-e3$	$2e5$
$e2$	0	0	0	$2e2$	$2e1$	$-2e3+4e4$
$e3$	0	0	0	0	0	0
$e4$	$-e1$	$-2e2$	0	0	$e5$	$2e6$
$e5$	$e3$	$-2e1$	0	$-e5$	0	0
$e6$	$-2e5$	$2e3-4e4$	0	$-2e6$	0	0

8. Maple Packages and Java Applets

Define a pair of vectors in the Lie algebra and compute the Lie bracket.

> `X:=evalV(e1 + 2*e4 -e6);`

$$X := e1 + 2\,e4 - e6$$

> `Y:=evalV(e2 -e3 +3*e5);`

$$Y := e2 - e3 + 3\,e5$$

> `Lie_bracket(X,Y);`

$$-4\,e2 - 5\,e3 + 4\,e4 + 6\,e5$$

Find the center of the Lie algebra.

> `center();`

$$[e3\,]$$

Compute the radical of the Lie algebra (the largest solvable ideal).

> `radical();`

$$[e3\,,\,e5\,,\,e1\,]$$

> `check_indecomposable();`

$$true$$

> `check_solvable();`

$$false$$

> `check_semi_simple();`

$$false$$

Since the algebra is indecomposable, not solvable and not semi-simple, the Levi decomposition is non-trivial and is now computed. The first factor is the radical and the second factor is the semi-simple part.

> `LD:=Levi_decomposition();`

$$LD := [[e5\,,\,e1\,,\,e3\,],\,[e2\,,\,(\frac{-1}{2})\ e3 + e4\,,\,e6\,]]$$

We can take any subalgebra of a Lie algebra and initialize it as a Lie algebra in its own right.

>   subalgebra_to_Lie_algebra_data(LD[2],SS);

    [[*Lie_alg*, *SS*, [3]], [[[1, 2, 1], 2], [[1, 3, 2], 4], [[2, 3, 3], 2]]]
>   Lie_alg_init(%);

Lie algebra : *SS*

The program classify_Lie_algebra will classify any 3 or 4 dimensional Lie algebra in accordance with a list of such algebras made by Pavel Winternitz. In this list the Lie algebra [3,5] denotes the algebra $sl(2)$. We conclude that this is the semi-simple part of the 6 dimensional Lie algebra of the 2-dimensional heat equation.

>   classify_Lie_algebra()[2];

[[*winternitz*, [3, 5]]]

References

[1] Anderson, I.M. (1992). *Introduction to the variational bicomplex*, Mathematical Aspects of Classical Field Theory, Contemporary Mathematics **132** (Amer. Math Soc., Providence) 51–73.
[2] Anderson, I.M. and Fels, M.E. (1997). Symmetry reduction of variational bicomplexes and the principle of symmetry criticality, *Amer. J. Math.* **112**, 609–670.
[3] Anderson, I.M., Fels, M.E. and Torre, C.G. (2000). Group invariant solutions without transversality, *Comm. Math. Physics* **212**, 653–686.
[4] Olver, P.J. (1986). *Applications of Lie Groups to Differential Equations*, 2nd ed. (Springer, New York).
[5] Petrov, A.J. (1969). *Einstein Spaces* (Pergamon, Oxford).